Current Controversies

# Climate Change and Biodiversity

# Other Books in the Current Controversies Series

America's Role in a Changing World
Artificial Intelligence and the Future of Humanity
Big Tech and Democracy
The Business of College Sports: Who Earns What
The Capitol Riot: Fragile Democracy
Cyberterrorism
Domestic Extremism
Media Trustworthiness
The Politics and Science of COVID-19
Reparations for Black Americans
Space: Tourism, Competition, Militarization

Current
CONTROVERSIES

# Climate Change and Biodiversity

**Lisa Idzikowski, Book Editor**

GREENHAVEN
PUBLISHING

Published in 2023 by Greenhaven Publishing, LLC
29 E. 21st Street
New York, NY 10010



First Edition



Articles in Greenhaven Publishing anthologies are often edited for length to meet page requirements. In addition, original titles of these works are changed to clearly present the main thesis and to explicitly indicate the author's opinion. Every effort is made to ensure that Greenhaven Publishing accurately reflects the original intent of the authors. Every effort has been made to trace the owners of the copyrighted material.

Cover image: Richard Whitcombe/Shutterstock.com

Library of Congress Cataloging-in-Publication Data

Names: Idzikowski, Lisa, editor.
Title: Climate change and biodiversity / Lisa Idzikowski, Book Editor.
Description: First edition. | New York : Greenhaven Publishing, 2023. | Series: Current controversies | Includes bibliographical references and index. | Audience: Ages 15+ | Audience: Grades 10–12 | Summary: "Anthology of diverse perspectives that explore the controversies surrounding climate change and biodiversity today"— Provided by publisher.
Identifiers: LCCN 2022001385 | ISBN 9781534508958 (library binding) | ISBN 9781534508941 (paperback) | ISBN 9781534508965 (ebook)
Subjects: LCSH: Biodiversity—Climatic factors—Juvenile literature. | Bioclimatology—Juvenile literature. | Climatic changes—Juvenile literature. | Nature—Effect of human beings on—Juvenile literature.
Classification: LCC QH543 .C555 2023 | DDC 577.2/2—dc23/eng/20220120
LC record available at https://lccn.loc.gov/2022001385

*Manufactured in the United States of America*

Website: http://greenhavenpublishing.com

# Contents

# Foreword

"Controversy" is a word that has an undeniably unpleasant connotation. It carries a definite negative charge. Controversy can spoil family gatherings, spread a chill around classroom and campus discussion, inflame public discourse, open raw civic wounds, and lead to the ouster of public officials. We often feel that controversy is almost akin to bad manners, a rude and shocking eruption of that which must not be spoken or thought of in polite, tightly guarded society. To avoid controversy, to quell controversy, is often seen as a public good, a victory for etiquette, perhaps even a moral or ethical imperative.

Yet the studious, deliberate avoidance of controversy is also a whitewashing, a denial, a death threat to democracy. It is a false sterilizing and sanitizing and superficial ordering of the messy, ragged, chaotic, at times ugly processes by which a healthy democracy identifies and confronts challenges, engages in passionate debate about appropriate approaches and solutions, and arrives at something like a consensus and a broadly accepted and supported way forward. Controversy is the megaphone, the speaker's corner, the public square through which the citizenry finds and uses its voice. Controversy is the life's blood of our democracy and absolutely essential to the vibrant health of our society.

Our present age is certainly no stranger to controversy. We are consumed by fierce debates about technology, privacy, political correctness, poverty, violence, crime and policing, guns, immigration, civil and human rights, terrorism, militarism, environmental protection, and gender and racial equality. Loudly competing voices are raised every day, shouting opposing opinions, putting forth competing agendas, and summoning starkly different visions of a utopian or dystopian future. Often these voices attempt to shout the others down; there is precious little listening and considering among the cacophonous din. Yet listening and considering, too, are essential to the health of a democracy. If controversy is democracy's lusty lifeblood, respectful listening and careful thought are its higher faculties, its brain, its conscience.

Current Controversies does not shy away from or attempt to hush the loudly competing voices. It seeks to provide readers with as wide and representative as possible a range of articulate voices on any given controversy of the day, separates each one out to allow it to be heard clearly and fairly, and encourages careful listening to each of these well-crafted, thoughtfully expressed opinions, supplied by some of today's leading academics, thinkers, analysts, politicians, policy makers, economists, activists, change agents, and advocates. Only after listening to a wide range of opinions on an issue, evaluating the strengths and weaknesses of each argument, assessing how well the facts and available evidence mesh with the stated opinions and conclusions, and thoughtfully and critically examining one's own beliefs and conscience can the reader begin to arrive at his or her own conclusions and articulate his or her own stance on the spotlighted controversy.

This process is facilitated and supported in each Current Controversies volume by an introduction and chapter overviews that provide readers with the essential context they need to begin engaging with the spotlighted controversies, with the debates surrounding them, and with their own perhaps shifting or nascent opinions on them. Chapters are organized around several key questions that are answered with diverse opinions representing all points on the political spectrum. In its content, organization, and methodology, readers are encouraged to determine the authors' point of view and purpose, interrogate and analyze the various arguments and their rhetoric and structure, evaluate the arguments' strengths and weaknesses, test their claims against available facts and evidence, judge the validity of the reasoning, and bring into clearer, sharper focus the reader's own beliefs and conclusions and how they may differ from or align with those in the collection or those of classmates.

Research has shown that reading comprehension skills improve dramatically when students are provided with compelling, intriguing, and relevant "discussable" texts. The subject matter of these collections could not be more compelling, intriguing, or urgently relevant to today's students and the world they are poised to inherit. The anthologized articles also provide the basis for stimulating, lively, and passionate classroom debates. Students who are compelled to anticipate objections to their own argument and identify the flaws in those of an opponent read more carefully, think more critically, and steep themselves in relevant context, facts, and information more thoroughly. In short, using discussable text of the kind provided by every single volume in the Current Controversies series encourages close reading, facilitates reading comprehension, fosters research, strengthens critical thinking, and greatly enlivens and energizes classroom discussion and participation. The entire learning process is deepened, extended, and strengthened.

If we are to foster a knowledgeable, responsible, active, and engaged citizenry, we must provide readers with the intellectual, interpretive, and critical-thinking tools and experience necessary to make sense of the world around them and of the all-important debates and arguments that inform it. We must encourage them not to run away from or attempt to quell controversy but to embrace it in a responsible, conscientious, and thoughtful way, to sharpen and strengthen their own informed opinions by listening to and critically analyzing those of others. This series encourages respectful engagement with and analysis of current controversies and competing opinions and fosters a resulting increase in the strength and rigor of one's own opinions and stances. As such, it helps readers assume their rightful place in the public square and provides them with the skills necessary to uphold their awesome responsibility—guaranteeing the continued and future health of a vital, vibrant, and free democracy.

# Introduction

> *"Mammals, fish, birds, reptiles, and other vertebrate species are disappearing many times faster than they should be, a phenomenon that has been linked to climate change, pollution, and deforestation—all interconnected threats."*
>
> *— Melissa Denchak and Jeff Turrentine*

No matter what part of the world a person calls home, they still occupy the only planet we all inhabit, Earth. From the first railroads, textile mills, Statue of Liberty, and factory farms in the United States, the Mountain Railways of India, the Eiffel Tower in France, coffee plantations in Brazil, palm oil conglomerates in Indonesia and Malaysia, and so many more imprints of modern civilization, human beings have left large footprints all over the natural world. It is this interaction with nature on Earth, especially since the Industrial Revolution, that plays an indelible role in climate change and biodiversity.

The topics of climate change and biodiversity loom large in today's consciousness. Would it be surprising to learn that climate change directly impacts biodiversity, while biodiversity can affect climate change? Biodiversity refers to the number and variety of species of plants and animals in an environment. When thinking about biodiversity, some people may picture giraffes and elephants roaming the African savanna, polar bears hunting for seals on sea ice, or even monarch butterflies and their milkweed host plants in nearby parks or backyard gardens. Whether it's eating a ripe banana, enjoying a favorite chocolate candy bar (both bananas and chocolate have issues concerning biodiversity), or watching a nature video about various animals and plants, people have some connection or experiential knowledge of biodiversity.

Familiarity with climate change is similar, though many confuse it with the term "global warming." Climate change is a long-term change of temperature and weather patterns. In the United States today, it is rare to find a news report that doesn't mention unusual weather. The deadly Texas cold snap of February 2021, the twenty-one storms of 2021's hurricane season, the all-time heat records of the US northwest in June 2021, or the smoke covering the middle and eastern US states that had belched from California's unprecedented wildfires starting that summer are just a few examples.

Surprisingly, while experts at NASA's Jet Propulsion Laboratory report that 97 percent or more of actively publishing climate scientists agree that climate change in the form of climate warming is due to human activities, only 54 percent of Americans believe that climate change is a result of human activity, and an unbelievable 10 percent don't believe in climate change at all, as reported in an October 2021 article from *Newsweek* online. On a global scale, 72 percent of people from a Pew Research Center survey conducted in spring 2021 were concerned that climate change would harm them personally at some point in their lifetime, and 52 percent were not confident that actions taken internationally would reduce the harmful effects of climate change globally.

If close to six in ten US adults think that effects from global warming are already happening as stated by Gallup, why does there seem to be so much confusion and non-action? Do people still confuse daily weather conditions with long-term climate? Or maybe they don't pay much attention to the weather at all? Perhaps people refuse to think that things they do personally contribute to the problem. Or maybe is it that people want to think that things will somehow magically improve? Hopefully, it's not because they don't care.

Evidence and explanations about climate change and its effects on biodiversity are all around. Scientific facts demonstrate that the changes on Earth have occurred substantially since the beginning of the Industrial Revolution and have increased greatly in just a few decades. And while Earth has gone through natural cycles of change over billions of years, nothing compares to the magnitude of changes in a short time, like the blink of an eye in geoscience years.

Where does this leave human society, where no one will ever again witness flocks of passenger pigeons so thick they blotted out the sky for days as they migrated overhead or hear the call of an ivory-billed woodpecker likened to a *toot-toot* sound from a child's toy horn? Actions can be taken on a global, governmental, or individual scale. At COP26, the United Nations Climate Change Conference of 2021, agreements between nations took big steps to address the crisis. The US and China pledged to cooperate more in switching to clean energy, and leaders from over 100 countries promised to end deforestation by 2030. But just as important are the steps that individuals can take to decrease their toll on the environment. These include decreasing consumption of meat and eliminating food waste, walking more than driving, and even putting up backyard houses for bats and bees.

Time is of the essence. It is not too late, but everyone must act. Listen to scientists, read reports from reliable sources, educate others, and believe that humanity can win the fight against climate change and biodiversity loss.

Begin by investigating and understanding the current debate surrounding this interesting topic. Read and contemplate the diverse viewpoints contained in *Current Controversies: Climate Change and Biodiversity*, which sheds light on this ongoing contemporary issue.

Current Controversies

# Chapter 1

# Does Climate Change Pose an Immediate Threat?

# Overview: What We Need to Know About Climate Change

*Melissa Denchak and Jeff Turrentine*

*Melissa Denchak is a freelance editor and writer. Jeff Turrentine is a columnist for the Natural Resources Defense Council.*

Climate change is generally defined as a significant variation of average weather conditions—say, conditions becoming warmer, wetter, or drier—over several decades or more. It's the longer-term trend that differentiates climate change from natural weather variability.

Record floods. Raging storms. Deadly heat. Climate change manifests itself in myriad ways and is experienced by every living being, although not equally. Throughout the world, the economically disadvantaged and people of color—those who have contributed the very least to the root causes of climate change—are the most likely to suffer from its worst impacts. Here are the basics on what causes climate change, how it's affecting the planet and its people, and what we can do about it.

## How Is Climate Change Measured Over Time?

Climate refers to the general weather conditions of a place as measured over many years. In the United States, for example, Maine's climate is cold and snowy in winter while South Florida's is tropical year-round.

Earth-orbiting satellites, remote meteorological stations, and ocean buoys are used to monitor present-day weather and climate, but it's paleoclimatology data from natural sources like ice cores, tree rings, corals, and ocean and lake sediments that have enabled scientists to extend the earth's climatic records back millions of

"Climate Change: What You Need to Know," by Melissa Denchak and Jeff Turrentine, NRDC (Natural Resources Defense Council), September 1, 2021. Reprinted by permission.

years. These records provide a comprehensive look at the long-term changes in the earth's atmosphere, oceans, land surfaces, and cryosphere (frozen water systems). Scientists then feed this data into sophisticated climate models that are able to predict future climate trends with impressive accuracy.

It's worth noting that while *climate change* and *global warming* are often used interchangeably, global warming—the recent rise in the global average temperature near the earth's surface—is just one aspect of climate change.

## What Causes Climate Change?

The mechanics of the earth's climate system are simple. When energy from the sun is reflected off the earth and back into space (mostly by clouds and ice), or when the earth's atmosphere releases energy, the planet cools. When the earth absorbs the sun's energy, or when atmospheric gases prevent heat released by the earth from radiating into space (the greenhouse effect), the planet warms. A variety of factors, both natural and human, can influence the earth's climate system.

### *Natural Causes of Climate Change*

The earth has gone through warming and cooling phases in the past, long before humans were around. Forces that can contribute to climate change include the sun's intensity, volcanic eruptions, and changes in naturally occurring greenhouse gas concentrations. But records indicate that today's climatic warming—particularly that which has occurred since the mid-20th century—is happening at a much faster rate than ever before, and it can't be explained by natural causes alone. According to NASA, "[t]hese natural causes are still in play today, but their influence is too small or they occur too slowly to explain the rapid warming seen in recent decades."

### *Anthropogenic Causes of Climate Change*

Humans—more specifically, the greenhouse gas (GHG) emissions that human activity generates—are the leading cause of the earth's rapidly changing climate today. Greenhouse gases play an important role in keeping the planet warm enough to inhabit. But the amount of these gases in our atmosphere has skyrocketed in recent decades. According to the U.S. Environmental Protection Agency, our current concentrations of carbon dioxide, methane, and nitrous oxide "are unprecedented compared with the past 800,000 years." Indeed, the atmosphere's share of carbon dioxide—the planet's chief climate change contributor—has risen by 46 percent since preindustrial times.

The burning of fossil fuels like coal, oil, and gas for electricity, heat, and transportation is the primary source of human-generated emissions. A second major source is deforestation, which releases sequestered (or stored) carbon into the air. It's estimated that logging, clearcutting, fires, and other forms of forest degradation release an average of 8.1 billion metric tons of carbon dioxide per year, accounting for more than 20 percent of all global CO2 emissions. Other human activities that generate air pollution include fertilizer use (a primary source of nitrous oxide emissions), livestock production (cattle, buffalo, sheep, and goats are major methane emitters), and certain industrial processes that release fluorinated gases. Activities like agriculture and road construction can also change the reflectivity of the earth's surface, leading to local warming or cooling.

Though our planet's forests and oceans absorb greenhouse gases from the atmosphere through photosynthesis and other processes, these natural carbon sinks can't keep up with our rising emissions. The resulting buildup of greenhouse gases is causing alarmingly fast warming worldwide. It's estimated that the earth's average temperature rose by about 1 degree Fahrenheit during the 20th century. If that doesn't sound like much, consider this: When the last ice age ended and the northeastern United States was covered by more than 3,000 feet of ice, average temperatures were just 5 to 9 degrees cooler than they are now.

## The Effects of Global Climate Change

According to the World Economic Forum's Global Risks Report 2021, the failure to mitigate and adapt to climate change is "the most impactful" risk facing communities worldwide—ahead of even weapons of mass destruction and water crises. Blame its cascading effects: As climate change transforms global ecosystems, it affects everything from the places we live to the water we drink to the air we breathe.

And though climate change affects everyone in some way, it's indisputable that its most negative impacts are borne disproportionately by certain groups: women, children, people of color, Indigenous communities, and the economically marginalized. Climate is a human rights issue.

### *Extreme Weather*

As the earth's atmosphere heats up, it collects, retains, and drops more water, changing weather patterns and making wet areas wetter and dry areas drier. Higher temperatures worsen and increase the frequency of many types of disasters, including storms, floods, heat waves, and droughts. These events can have devastating and costly consequences, jeopardizing access to clean drinking water, fueling out-of-control wildfires, damaging property, creating hazardous-material spills, polluting the air, and leading to loss of life.

### *Dirty Air*

Air pollution and climate change are inextricably linked, with one exacerbating the other. When the earth's temperatures rise, not only does our air get dirtier—with smog and soot levels rising accordingly—but it also becomes filled with more allergenic pollutants, such as circulating mold (thanks to damp conditions from extreme weather and more floods) and pollen (due to longer, stronger pollen seasons).

### *Health Risks*

According to the World Health Organization, "climate change is expected to cause approximately 250,000 additional deaths per year" between 2030 and 2050. As global temperatures rise, so do the number of fatalities and illnesses from heat stress, heatstroke, and cardiovascular and kidney disease. And as air pollution worsens, so does respiratory health—particularly for the 300 million people living with asthma worldwide; there's more airborne pollen and mold to torment hay fever and allergy sufferers too. Extreme weather events, such as severe storms and flooding, can lead to injury, drinking water contamination, and storm damage that may compromise basic infrastructure or lead to community displacement. Indeed, historical models suggest the likelihood of being displaced by a disaster is now 60 percent higher than it was four decades ago—and the largest increases in displacement are being driven by weather- and climate-related events. (It's worth noting that displacement comes with its own health threats, such as increases in urban crowding, trauma, social unrest, lack of clean water, and transmission of infectious diseases.) A warmer, wetter world is also a boon for insect-borne diseases such as dengue fever, West Nile virus, and Lyme disease.

### *Rising Seas*

The Arctic is heating up twice as fast as any other place on the planet. As its ice sheets melt into the seas, our oceans are on track to rise by anywhere from 0.95 to 3.61 feet by the end of this century, threatening coastal ecosystems and low-lying areas. Island nations face particular risk, as do some of the world's largest cities, including New York City, Miami, Mumbai in India, and Sydney in Australia.

### *Warmer, More Acidic Oceans*

The earth's oceans absorb between one-quarter and one-third of our fossil fuel emissions and are now 30 percent more acidic than they were in preindustrial times. This acidification poses a serious threat to underwater life, particularly creatures with calcified shells or skeletons like oysters, clams, and coral. It can have a devastating impact on shellfisheries, as well as the fish, birds, and mammals that depend on shellfish for sustenance. In coastal communities where fishing and seafood production sustain the local economy, this impact extends to human populations as well, destroying livelihoods and opening the door to economic ruin. Rising ocean temperatures are also altering the range and population of underwater species and contributing to coral bleaching events that are capable of killing entire reefs—ecosystems that support more than 25 percent of all marine life.

### *Imperiled Ecosystems*

Climate change is increasing pressure on wildlife to adapt to changing habitats—and fast. Many species are seeking out cooler climates and higher altitudes, altering seasonal behaviors, and adjusting traditional migration patterns. These shifts can fundamentally transform entire ecosystems and the intricate webs of life that depend on them. As a result, according to a 2020 study, one-third of all animal and plant species could face extinction by 2070. Another study showed that mammals, fish, birds, reptiles, and other vertebrate species are disappearing many times faster than they should be, a phenomenon that has been linked to climate change, pollution, and deforestation—all interconnected threats. In early 2021, climate and biodiversity experts issued a joint report illustrating these interconnections and urging simultaneous action on both fronts. On the flip side, milder winters and longer summers have enabled some species to thrive, including tree-killing insects that are endangering entire forests.

## Climate Change Facts

Despite what climate deniers and fossil fuel lackeys claim—for instance, that the science on global warming is "far from settled"—there's nothing to debate: Climate change is a reality. In the words of the Intergovernmental Panel on Climate Change (IPCC), the foremost international scientific body for the assessment of the phenomenon, "[w]arming of the climate system is unequivocal, and since the 1950s, many of the observed changes are unprecedented over decades to millennia. The atmosphere and ocean have warmed, the amounts of snow and ice have diminished, and sea level has risen."

But you don't have to be a climate scientist to understand how the world has already changed. Our last decade—2010 to 2019—was hotter than any other decade in at least the past 1,300 years. Second place goes to the decade before it: 2000 to 2009. And the first year of the new decade, 2020, is tied with 2016 for the hottest individual year ever recorded. With this increased heat, of course, have come more frequent and more intense weather-related disasters. The first half of 2021 was marked by record heat, drought, and wildfires in the western United States and Canada and by record rainfall and flooding in Europe. And without immediate climate action, these records won't be records for long: They'll be broken next year, or shortly thereafter. "The science is clear," writes Vijay Limaye, NRDC scientist. "These dangerous effects of climate change will worsen each year that we fail to curb the pollution that is destabilizing our planet."

The responsibility to reverse this worrying trend lies with us. At least 97 percent of actively publishing climate scientists endorse the consensus position that humans are the lead drivers of climate change. As the IPCC has stated—with its highest degree of confidence—"It is extremely likely that more than half of the observed increase in global average surface temperature from 1951 to 2010 was caused by the anthropogenic increase in GHG concentrations and other anthropogenic forcings together."

## Climate Change Solutions

We can mitigate global climate change and help stem its detrimental impacts, but doing so will require tackling its root cause: pollution from burning fossil fuels.

### *Paris Agreement*

The Biden administration has committed to the principles set forth during the 2015 Paris Climate Change Conference, where nearly every nation on the earth committed to actions aimed at shifting away from dirty fossil fuels and toward cleaner, smarter energy options in order to limit global temperature rise this century to 2 degrees Celsius—or 1.5 degrees Celsius, if possible.

On Earth Day in 2021, President Biden committed the United States to cutting its carbon pollution by 50 to 52 percent from 2005 levels—nearly double what President Obama had promised—and to do it by 2030, just five years later than the original deadline set when the United States first joined the agreement. While bold, the new targets are achievable, but only if Congress, states, and the business sector join in to help us get there by doing things like setting a national clean energy standard, electrifying our fleet of cars and trucks, curbing methane emissions, and maximizing building efficiency.

### *Developments in Clean Energy Technology*

Every year brings with it news of exciting breakthroughs in the production and distribution of renewable energy sources such as wind and solar, which are becoming ever cheaper to generate and ever more popular with consumers, businesses, utilities, and governments. At the same time, carmakers around the world are working to develop new battery technologies that will result in the replacement of hundreds of millions of gas-powered cars with zero-emissions vehicles. These trends should be encouraged by smart, forward-looking public policy—such as Biden's Build Back Better agenda, which ties the post-pandemic economic recovery in the United States to projects that will curb climate change while putting tens of millions of unemployed or underemployed people to work in good-paying jobs.

### *Climate Action at Home*

Tackling global climate change is a Herculean task, one that depends on international consensus and the efforts of communities, companies, and individuals alike. To that end, many states, from California to Iowa, are championing clean energy industries, such as solar and wind; U.S. cities are taking action to mitigate climate change and bolster climate resilience while placing equity at the forefront; and corporations—including some of the world's largest multinationals—are pledging to change the way they do business to achieve net-zero emissions by 2040.

There are myriad ways that you can help too. You can join, amplify, or otherwise support the international youth climate movement that has emerged in the last few years as a galvanized response to government inaction; in rallies and marches around the world, this movement is letting leaders know that delay on climate action will no longer be tolerated. You can also pick up the phone to call Congress about environmental policies that matter, supporting renewable energy projects, and prioritizing fuel and energy efficiency that will not only curb individual carbon emissions but bolster clean alternatives to dirty fossil fuels. We must all step up—and now.

# Climate Change Is Affecting Life in the Arctic

***Michael R. Heath et al.***

*Michael Heath has been studying marine ecosystems for forty years. His current research uses mathematical statistics to investigate the dynamics of ecosystems. Heath's co-researchers on the project are invested in marine ecology.*

The poles are among the harshest environments on our planet. In the winter, the sun never rises, air temperatures are extremely cold, and snow and ice accumulate. In summer, the sun never sets, it is warmer, and the ice melts back toward the poles. However, there is no doubt that the amount of ice is shrinking as our planet gets warmer. This is happening on the land and at sea. On the land, glaciers are melting and retreating in places like Antarctica, Greenland, and Alaska. At sea, the area of floating ice, which we call sea-ice, is becoming smaller each year.[1]

A National Geographic video of a starving polar bear with the caption "This is what climate change looks like" has become an internationally recognizable symbol of the effects of melting sea-ice on the Arctic ecosystem.[2, 3] The story behind the video is that polar bears depend on being able to roam over the ice in search of seals to eat. Because the ice is shrinking they are struggling to find enough food. In this article, we explain why the story may be more complex than it first appears, and how computer models of ecosystems can be used to help us understand these complexities.

## Food Webs and Ecosystems

An ecosystem is a community of bacteria, plants, and animals that occupy the same habitat. All the organisms in an ecosystem

Originally published as: "How Is Climate Change Affecting Marine Life in the Arctic?" by Michael R. Heath, Deborah Benkort, Andew S. Brierley, Ute Daewel, Richard Hofmeister, Jack H. Laverick, Roland Proud and Douglas C. Speirs, Frontiers Media S.A, 08/14/2020. https://kids.frontiersin.org/articles/10.3389/frym.2020.00103. 

are interconnected by a network of who-eats-whom relationships. We call this a food web. Wherever there is sunlight and sufficient nutrients, plants manufacture the materials needed for life in a process called photosynthesis. Photosynthesis provides the energy to support all the animals in the rest of the food web.

In the sea, most of the plants are microscopic and we call them phytoplankton. Phytoplankton are eaten by a variety of tiny animals up to about the size of a grain of rice, called zooplankton. Zooplankton are then eaten by larger zooplankton, jellyfish, and plankton-eating fish, such as capelin. The waste they produce (called detritus) is broken down by bacteria as it falls to the seabed, where it provides food for bottom-living animals (called benthos). Plankton-eating fish and benthos are food for predatory fish, such as cod and hake, seabirds, and mammals, such as seals and whales. In the Arctic, all these animals are hunted by the top predators, such as polar bears. Humans are also top predators in the food web, because we catch fish and hunt for seals and whales.

## How Is Warming of the Planet Affecting Arctic Food Webs?

Warming has a direct effect on all living organisms. Metabolism depends on temperature.[4] Metabolism includes the processes that allow organisms to produce energy from the food that they eat. All organisms have a temperature range in which their metabolism can keep them living comfortably. Tropical species are less tolerant of cold conditions than polar species. This means that, as oceans warm, we expect species found in warm regions to expand their distributions, and we expect those species that prefer colder water to retreat toward the poles. Over the past 50 years, scientists have observed some zooplankton and fish species moving toward the poles, as the oceans become warmer.[5]

The effects of warming on the sea-ice are also extremely important for the Arctic food webs. The presence of sea-ice strongly reduces the amount of sunlight that enters the ocean. The shrinking area of sea-ice means that some parts of the ocean are being exposed to summer sunlight for the first time in many thousands of years. The light enables phytoplankton to grow and photosynthesize in areas that have previously been barren, increasing the productivity of the Arctic.

The increase in phytoplankton could be considered a positive change, leading to more life in the sea and more food for zooplankton, benthos, and eventually fish, birds, seals, whales, and polar bears.[6] However, some of the Arctic species have evolved to depend on the presence of ice. Sea-ice provides a means for seals and walruses to get out of the water. They need to do this to rest, breed, and evade other marine predators, such as killer whales. Other mammals, such as polar bears, rely on the ability to travel across the frozen sea to hunt for seals. Melting ice makes travel more difficult because thin ice cannot support the bears' weight. On the other hand, ice is a barrier to whale species because they must be able to reach the air to breathe. Overall, it is not completely clear which of the predators in the food web will be benefited by melting sea-ice and which will be harmed by it.

## Predicting What Will Happen in the Arctic

We use computer models to make predictions about the effects that changes in sea-ice will have on food webs. To make these models, we first write down the key processes and components of the food web as mathematical equations. These include feeding, metabolism, photosynthesis, and who-eats-whom. Then we convert the equations into computer code. This gives us a virtual world, in which we can carry out experiments on the model ecosystem. The experiments are designed to help us understand how the ecosystem could change in the future.

We built a simple model of the Barents Sea ecosystem in the Norwegian/Russian Arctic. This model contains data on ice and temperature conditions, and equations to represent some of what we know about the ecosystem's plants and animals. The main simplification is that some species are combined into groups that share similar characteristics, such as size, structure, or diet. For example, we combined all birds and mammals into a single group, because they are warm-blooded, which means that they can use energy from their food to maintain a warm body temperature independent of the environment, unlike plankton and most fish. This simplification means that, in our model, we cannot distinguish between polar bears and whales, for example. Although our simplified model is not perfect, it still reproduces the general responses of the natural Barents Sea ecosystem.

The equations in the model include parameters that determine the results. In our model, these include, for example, maximum feeding rates, or sinking rates of detritus. Selecting values for the parameters is an important step in building the model. We try to select values that allow the model to match as closely as possible to the real data from the ecosystem. This ensures that the model is as realistic as possible. Once the model is set up, we can change the ice and temperature conditions to what they might be in the future, and use the model to predict what may happen in real life. In a comparison between the present day and future model results, the present-day model represents the Barents Sea as it is today. The future model is ice-free, with water temperatures 0.5°C warmer all year round. As expected, photosynthesis is higher in the future model. The animals that benefit are benthos, predatory fish, and the bird-and-mammal group. Zooplankton and plankton-eating fish are less abundant than they are today. This is because phytoplankton and detritus, upon which the benthos and zooplankton feed, settle to the seabed more quickly when there is no ice. Since we cannot distinguish between different bird and mammal predators in this model, we cannot use it to say exactly which species will be more abundant or less abundant in the future. However, the model predicts that the bird-and-mammal group as a whole will have a diet different from today's, with more feeding on predatory fish and less on planktivorous fish.

## Why Does All This Matter?

The Arctic is warming faster than the rest of the planet. Yearly averaged air temperatures in the Arctic have increased by around 2°C between 1970 and 2010 compared to 0.6°C for the planet as a whole.[7] Summer temperatures in the Canadian Arctic are now higher than at any time in the last 44,000 years.[8] Within the period of a human lifespan, the Arctic Ocean has partially emerged from beneath thousands of years of sea-ice cover.

There is a race between nations to claim the seabed and resources in the Arctic as it emerges from the ice. New routes will become possible for ships to travel between Europe, Asia, and North America. On 30 November 2017, the EU and nine major fishing nations agreed not to develop fisheries in the Arctic Ocean until at least 2033. This is supposed to allow time for scientists to develop the models needed for managing the fish stocks in a sustainable way.[9] There is much research to be completed before we have a full understanding of the effects of warming on the Arctic.

### *References*

1. Comiso, J. C. 2012. Large decadal decline of the arctic multiyear ice cover. *J. Clim.* 25:1176–93. doi: 10.1175/JCLI-D-11-00113.1

2. Gibbens, S. 2017. *Heart-Wrenching Video Shows Starving Polar Bear on Iceless Land.* Available online at: https://www.nationalgeographic.com/news/2017/12/polar-bear-starving-arctic-sea-ice-melt-climate-change-spd/

3. Mittermeier, C. G. 2018. *Starving-Polar-Bear Photographer Recalls What Went Wrong.* Available online at: www.nationalgeographic.com/magazine/2018/08/explore-through-the-lens-starving-polar-bear-photo/

4. Gillooly, J. F., Brown, J. H., West, G. B., Savage, V. M., and Charnov, E. L. 2001. Effects of size and temperature on metabolic rate. *Science* 293:2248–51. doi: 10.1126/science.1061967

5. Kortsch, S., Primicerio, R., Fossheim, M., Dolgov, A. V., and Aschan, M. 2015. Climate change alters the structure of arctic marine food webs due to poleward shifts of boreal generalists. *Proc. R. Soc. B Biol. Sci.* 282:20151546. doi: 10.1098/rspb.2015.1546

6. Darnis, G., Robert, D., Pomerleau, C., Link, H., Archambault, P., Nelson, R. J., et al. 2012. Current state and trends in Canadian Arctic marine ecosystems: II. Heterotrophic food web, pelagic-benthic coupling, and biodiversity. *Clim. Change* 115:179–205. doi: 10.1007/s10584-012-0483-8

7. IPCC. 2013. Climate Change 2013: The Physical Science Basis. Contribution of Working Group I to the Fifth Assessment Report of the Intergovernmental Panel on Climate Change. Cambridge; New York, NY: Cambridge University Press.

8. Miller, G. H., Lehman, J. J., Refsnider, K. A., Southon, J. R., and Zhong, Y. 2013. Unprecedented recent summer warmth in Arctic Canada. *Geophys. Res. Lett.* 40:5745–51. doi: 10.1002/2013GL057188

9. Hoag, H. 2017. Nations agree to ban fishing in Arctic Ocean for at least 16 years. *Science*. doi: 10.1126/science.aar6437

## Animals and Plants Cannot Survive in Increasingly Intense Fires

*Cris Brack*

*Cris Brack is an associate professor at the Fenner School of Environment in Australia.*

Australia is a land that has known fire. Our diverse plant and animal species have become accustomed to life with fire, and in fact some require it to procreate.

But in recent decades the pattern of fires—also known as the fire regime—is changing. Individual fires are increasingly hotter, more frequent, happening earlier in the season and covering larger areas with a uniform intensity. And these changes to the fire regime are occurring too fast for our native flora and fauna to adapt and survive.

### Our Fire-Adapted Plants Are Suffering

Many of Australia's iconic eucalypts are "shade intolerant" species that adapted to exist within a relatively harsh fire regime. These species thrive just after a major fire has cleared away the overstory and prepared an ash bed for their seeds to germinate.

Some of our most majestic trees, like the alpine ash, can only regenerate from seed. Those seeds germinate only on bare earth, where the leaf litter and shrubs have been burnt away.

But if fire is so frequent the trees haven't matured enough to produce seed, or so intense it destroys the seeds present in the canopy and the ground, then even these fire-adapted species can fail.

The current fires are re-burning some forests that were burnt only a decade ago. Those regenerating trees are too young to survive, but also too young to have started developing seed.

With the disappearance of these tree species, other plants will fill the gap. Acacias (wattles) are potential successors as they mature much earlier than alpine ash. Our tall, majestic forests could easily turn into shrubby bushland with more frequent fires.

Even within a burnt area, there are usually some unburnt patches, which are highly valuable for many types of plants and animals. These patches include gullies and depressions, but sometimes are just lucky coincidences of the terrain and weather. The patches act as reserves of "seed trees" to provide regeneration opportunities.

Recent fires, burning in hotter and drier conditions, are tending to be severe over large areas with fewer unburnt patches. Without these patches, there are no trees in the fire zone to spread seeds for regeneration.

Eucalypt seed is small and without wings or other mechanisms to help the wind disperse it. Birds don't generally disperse these seeds either. Eucalypt seed thus only falls within 100–200 metres of the parent tree. It may take many decades for trees to recolonise a large burnt area.

That means wind-blown or bird-dispersed seeds from other species may fully colonise the burnt area well before the Eucalypts. Unfortunately many of these windblown seeds will be weed species, such as African Love Grass, which may then cover the bare earth and exclude successful Eucalypt regeneration while potentially making fires even hotter and more frequent.

## Animals Have Fewer Places to Hide

Young animals are significantly more vulnerable to disturbances such as fire than mature individuals. So the best time to give birth is a season when fire is rare.

Spring in the southern zones of Australia has, in the past, been wetter and largely free from highly destructive fires. Both flora and fauna species thus time their reproduction for this period. But as fire seasons lengthen and begin earlier in the year, vulnerable nestlings and babies die where they shelter or starve as the fires burn the fruits and seeds they eat.

Australian fauna have developed behaviours that help them survive fire, including moving towards gullies and depressions, climbing higher, or occupying hollows and burrows (even if not their own) when they sense fire.

But even these behaviours will fail if those refuges are uncharacteristically burning under hotter and drier conditions. Rainforest, marshes and the banks of watercourses were once safe refuges against fire, but we have seen these all burn in recent fires.

## What Can Be Done?

All aspects of fire regimes in Australia are clearly changing as a result of our heating and drying climate. But humans can have a deliberate effect, and have done so in the past.

Indigenous burning created a patchwork of burnt areas and impacted on the magnitude and frequency of fires over the landscape. These regular burns kept the understory under control, while the moderate intensity and patchiness allowed larger trees to survive.

There have been repeated calls of late to reintroduce Indigenous burning practices in Australia. But this would be difficult over vast areas. It requires knowledgeable individuals to regularly walk through each forest to understand the forest dynamics at a very fine scale.

More importantly, our landscapes are now filled with dry fuel, and shrubs that act as "ladders"—quickly sending any fire into tree canopies to cause very destructive crown fires. Given these high fuel conditions along with their potentially dangerous distribution, there may be relatively few safe areas to reintroduce Indigenous burning.

The changed fire conditions still require active management of forests, with trained professionals on the ground. Refuges could be developed throughout forests to provide places where animals can shelter and from which trees can recolonise. Such refuges could be reintroduced by reducing forest biomass (or fuel) using small fires where feasible or by mechanical means.

Biomass collected by machines could be used to produce biochar or other useful products. Biochar could even be used to improve the soil damaged by the fires and excess ash.

Midstory species could be cut down to prevent the development of fire ladders to tree crowns. Even the overstory could be thinned to minimise the potential for crown fires. Seed could also be collected from thinned trees to provide an off-site bank as ecological insurance.

Such active management will not be cheap. But using machinery rather than fire could control biomass quantity and distribution in a much more precise way: leaving some biomass on the ground as habitat for insects and reptiles, and removing other patches to create safer refuges from the fires that will continue to come.

# Climate Change Affects Animals, Plants, and People

### *Greentumble*

*Greentumble was founded to inspire others toward a sustainable lifestyle for the benefit of the environment.*

We often consider climate change in terms of how it will directly affect the human population—for example, greater extremes of temperature, more frequent droughts and flooding, and sea level rise impacting on our cities, homes, and the global economy.

But arguably the most significant effects of a warming planet and changing weather patterns will be on other living species; plants, animals, birds and marine life. Biodiversity—the number and range of living species—is expected to suffer a significant decline as the climate warms, resulting in negative impacts on global ecosystems and, ultimately, on humankind.

## Impacts of Climate Change on Biodiversity

### *#1 Effects of Rising Temperature*

Current international efforts to restrict climate change to less than 2 degrees Celsius might suggest that the expected temperature change is not huge, and therefore little damage will be done. There are two important factors, however, that mean that even a modest rise in the average global temperature will result in irreversible harm to living organisms and ecosystems.

First, the effects of global warming are not uniform across the planet's surface.

Some areas have already begun to experience abnormal weather patterns, such as extended periods of extreme temperatures or

"How Does Climate Change Affect Biodiversity?" Greentumble, August 6, 2018. https://greentumble.com/how-does-climate-change-affect-biodiversity. Reprinted by permission.

lower precipitation, resulting in the expansion of desert areas, shrinkage of glaciers, and death of coral reefs.

Changes to global weather patterns and ocean currents mean that the effects will be localized, often disproportionately affecting the most sensitive habitats.

The Arctic, Antarctic and high latitudes have experienced the highest rates of warming, a trend that is projected to continue.[1]

Second, the increase in temperature is happening more quickly than the rate at which most living species can adapt.

Many living organisms are exceptionally good at adapting to new environments, and have done so as the Earth's climate has changed in past millennia; for most, however, the speed of anthropogenic climate change is simply too fast for evolution to keep up.

Those that cannot adapt quickly enough are at risk of extinction.

### *#2 Changes in the Population and Distribution of Species*

One of the recurring images used to illustrate climate change is that of the polar bear. Although it has become a cliché, it does demonstrate one of the major effects of climate change for large mammals; loss of habitat.

Like many other species that have adapted to live on or under the ice, animals such as polar bears rely on low temperatures to maintain the sea ice they need to travel, hunt and breed successfully.

Ice-free summers in the Arctic would mean the loss of the polar bears' entire habitat and the destruction of a biome that supports an entire food chain including algae, invertebrates, birds, fish and marine mammals. As the top predator, polar bears eat seals, which in turn eat Arctic cod. The cod feed on zooplankton, which themselves consume algae, which inhabit ice pockets.

However, if rising temperatures cause ice to melt, then the algae will lose their habitat, leaving the zooplankton short of their preferred food and with the ripple effect running all the way up to the polar bears. So, rapidly shrinking icecaps will leave insufficient habitat for these populations to survive, resulting in large scale loss of biodiversity.

In other regions, species have begun to alter or expand the territories they cover in response to increasing temperatures. This has resulted in a shift to cooler and moister regions, to higher elevations in mountainous areas, and towards the poles, in order to stay within a similar climate range.

For example, 37 species of dragonfly have moved 30 kilometers northwards per decade in the UK, while several species of small mammal have shifted 50 meters per decade upwards in Yosemite National Park.

For mammals, birds, insects and invertebrates, this shift is not simply because they prefer a cooler temperature. As the plants and smaller animals they rely on for food change their pattern of distribution, so the entire food chain must follow. When these interdependent species get "out of sync," the whole ecosystem may begin to collapse.

Furthermore, although population shifts may be beneficial in the short term, they are unsustainable over time as species reach geographical limits such as oceans or deserts, or manmade boundaries such as cities, or as the available habitat shrinks.

### *#3 Changing Seasons, Migration and Breeding Cycles*

Most species depend on seasonal cues to begin activities such as hibernation, migration, germination, flowering, nest-making and mating. As weather patterns change and the climate warms, the seasons may begin and end at different times, affecting the habits of some species.

When these natural cycles—known as phenology—are altered, the ecosystem can begin to malfunction. If the changes to different species of plants, insects, birds and mammals no longer match, then the food sources they rely on may not be available at crucial times, such as when chicks hatch.

To take a simple example, many birds time their nesting, mating and egg-laying so that the chicks will hatch at a time of maximum abundance, when there are plentiful caterpillars and others food sources available to feed them.

In turn, the moths and butterflies have timed their egg-laying and hatching so the young caterpillars can eat the new, tender leaves of their favorite food source.

Most plant species germinate and grow most vigorously when the weather is both warm and wet enough to sustain it: If changes to the weather pattern mean that a specific plant species germinates and completes its lifecycle too early, or grows poorly due to excessive heat or lack of rainfall, the caterpillars that depend on it will starve and the young birds may fail to thrive.

In ecosystems around the world, there are many thousands of interdependent species—from microbes in the soil to top predators such as wolves and bears—that have evolved together and rely entirely on the presence of other species in order to survive. If any one of these is unable to adapt to the effects of climate change, the consequences could be catastrophic.

### *#4 Introduction of New Diseases, Pathogens and Parasites*

As species move to take advantage of new habitats, the risk of diseases, pathogens and parasites spreading also increases and this can have serious ramifications for human health, fisheries and agriculture.

For example, the oyster parasite, *Perkinsus marinus*, can cause serious die-offs in oysters. As a result of climate change, the parasite has extended its range by around 483 kilometers (300 miles), from Chesapeake Bay to Maine, creating problems for local fisheries that rely on oysters for their livelihoods.

### *#5 The Rise of Invasive Species*

Of course, climate change is beneficial for some species of plants and animals. When conditions begin to change, the species that can adapt most quickly may begin to dominate an ecosystem, further threatening other, weaker species that respond more slowly, or replacing those that fall prey to an increase in pests or diseases.

Alien invasive species—those that did not evolve alongside native plants and animals but were introduced from elsewhere, deliberately or accidentally—are, by nature, highly adaptable and opportunistic. As climate change leads to extended growing seasons in some areas, invasive plants take advantage by flowering earlier and crowding out less adaptable indigenous species.

Since the native plants are much more likely to provide food, shelter and nesting material for native wildlife, dominant invasive species can quickly threaten an entire ecosystem and its biodiversity.

### *#6 Ocean Warming and Acidification*

Most of the warming that's happening right now is not on the earth's surface or in the atmosphere; it's in the oceans,[2] and that's where many ecosystem changes are also occurring.

Changing ocean temperatures, acidification, and circulation currents—all caused by greenhouse gas emissions—are likely to lead to irreversible ecological damage including extinctions and habitat loss.

As the concentration of carbon dioxide (CO2) in the atmosphere increases, more CO2 is absorbed by the ocean, which makes it more acidic. The higher level of acidity prevents marine organisms from making shells and skeletons from calcium carbonate, and can even dissolve shells that have already been formed. Many of the affected species play a vital role in the food chain and in ecosystem biodiversity.

Coral reefs, for example, contain around one third of the world's entire marine biodiversity and provide a range of ecosystem functions, but they are severely threatened by coral bleaching due to increased sea surface temperatures and acidification.

Researchers estimate that most coral reefs will suffer long-term degradation by 2030 even if we limit global warming to 2 degrees Celsius.[3]

## The Consequences of Biodiversity Loss

The most obvious consequence of biodiversity loss is large scale extinction of species, meaning that we would no longer have the great variety of wildlife and landscapes that we see and enjoy today. It is sad to think that in only a few generations' time, our descendants may not be able to see living examples of plants, animals and marine life we regard as commonplace.

Although we know that, over the history of the planet, many species have become extinct, nature is able to compensate for the loss of a small number of species over a long period of time, as other species evolve.

The rate of extinction is up to one thousand times higher than that seen in fossil records and the forecast extinction rate is more than ten times the current rate.[4]

Just as importantly, this rate of extinction and declining biodiversity will lead to the loss of a range of ecosystem services such as purification of air and water, climate stabilization, the generation of soil and natural vegetation and stable climatic conditions.

## How Will Climate Change and Biodiversity Loss Affect Us?

Human civilization relies on ecosystem functions to deliver a wide range of goods and services that are vital to support the continuation of human existence.

Loss of biodiversity affects the ability of ecosystems to provide us with food, clean air and fresh water, as well as medicines and genetic material—including many we may not even have discovered yet.

Because the many species that make up any ecosystem are interdependent, it's hard to predict exactly what the effects of reduced biodiversity in any one area might be.

One of the major global threats to human existence, however, is a decline in pollinating insects. Because different types of pollinator prefer different plant species, biodiversity ensures we can continue to enjoy many types of food. Loss of biodiversity means we may no longer be able to produce affordable honey, berries, nuts, coffee or chocolate, and if all pollinators disappeared we may lose up to one third of all crops worldwide,[5] resulting in widespread food shortages and famine.

Even coral reefs provide vital services, helping protect coastlines from the damaging effects of waves and storms, as well as providing food sources (including around 25 percent of the fish caught by developing nations[6]) and generating tourism. It has been estimated that the global value of coral reef ecosystem services is more than US$30 billion every year.[7]

Ultimately, we rely on the biodiversity of the many different ecosystems around us to make the planet habitable, provide for our most basic human needs and enhance our quality of life.

## How Can You Help to Protect Biodiversity?

As well as climate change, biodiversity is under threat from a number of human activities including pollution, over-exploitation of resources, urbanization and population growth.

Climate change, however, is something we can all help tackle every day.

Start now by aiming to:

- Cut your personal consumption of energy, water and other resources—at home, at school or at work.
- Reduce the amount of waste you produce, avoid disposable products, and recycle and compost your waste when possible.
- Buy locally made products, produced by people and organisations that you know are environmentally responsible.
- Walk, cycle or use public transport (or an electric vehicle, if you must drive).
- Learn more about your carbon footprint, and how you can minimize your climate change impact to help preserve biodiversity.

## *References*


1. Source: Global Biodiversity Outlook 3
2. http://www.climateshifts.org/?p=5585
3. https://www.pik-potsdam.de/news/press-releases/archive/2012/most-coral-reefs-are-at-risk-unless-climate-change-is-drastically-limited
4. Source: Millennium Ecosystem Assessment
5. https://www.nytimes.com/2016/02/27/science/decline-of-species-that-pollinate-poses-a-threat-to-global-food-supply-report-warns.html?_r=0
6. http://www.wmo.int/pages/prog/wcp/agm/publications/documents/Climate_Carbon_CoralReefs.pdf
7. http://www.wmo.int/pages/prog/wcp/agm/publications/documents/Climate_Carbon_CoralReefs.pdf

# Hurricanes Are Not Getting Worse Because of Climate Change

***Jon Miltimore***

*Jon Miltimore is the managing editor of FEE.org. His writing/reporting has been the subject of articles in* Time, *the* Wall Street Journal, *CNN,* Forbes, *Fox News, and the* Star Tribune.

In the wake of the destruction of Hurricane Ida, President Joe Biden this week traveled to storm-ravaged areas of New Jersey and New York to deliver a "code red" climate change message to the world: extreme weather poses an "existential threat" to humanity.

"The threat is here. It's not going to get any better. The question is can it get worse? We can stop it from getting worse," Biden said in the New York City borough of Queens, where he met with Senate Majority Leader Chuck Schumer, Gov. Kathy Hochul, Mayor Bill deBlasio and others. "This is everybody's crisis."

Biden was echoing what has essentially become conventional wisdom: climate change is making extreme weather much worse.

"Climate change has turbocharged severe storms, fires, hurricanes, coastal storms and floods—threatening millions," the *Washington Post* recently reported. "Nearly 1 in 3 Americans experienced a weather disaster this summer."

It's a theme routinely trotted out after hurricanes. Following Hurricane Katrina, a devastating Category 5 hurricane that caused more than 1,800 deaths and some $125 billion in damage in 2006, research claimed Atlantic hurricanes doubled in the last century.

"These numbers are a strong indication that climate change is a major factor in the increasing number of Atlantic hurricanes," said Greg Holland of the National Center for Atmospheric Research (NCAR) in Boulder, Colorado.

"The Myth That Hurricanes Are Getting Worse (Because of Climate Change)," by Jon Miltimore, Foundation for Economic Education, September 13, 2021. https://fee.org/articles/the-myth-that-hurricanes-are-getting-worse-because-of-climate-change/.

The idea that humanity is beset by an increased number of "turbocharged" storms is a bit frightening. But is it true?

## A Closer Look at the Science

Before you take out a loan to build a storm shelter in your basement, it might be worthwhile to look at data from the American Meteorological Society recently published in the *Wall Street Journal*. The data show fewer hurricanes are landing on the continental US, not more.

"[D]espite what you may have heard, Atlantic hurricanes are not becoming more frequent," explains Danish economist Bjorn Lomborg in the *Journal*. "In fact, the frequency of hurricanes making landfall in the continental U.S. has declined slightly since 1900."

The *WSJ* is a respected publication, but it of course has a reputation for being right of center. So it's important to note that Lomborg and the *Journal* are not out on a limb on this one. There is widespread consensus that hurricanes are not increasing in frequency.

"[A] new statistical analysis of historical records and satellite data suggests that there aren't actually more Atlantic hurricanes now than there were roughly 150 years ago, researchers report July 13 in Nature Communications," reported *Science News*.

The findings reported in Nature Communications were not an outlier. As *The Economist* reported in 2017 and the *Washington Post* reported in 2015, a plethora of research shows hurricanes are becoming less frequent, not more frequent.

That is only half of the story, however. While there is general agreement today that global warming is not causing more hurricanes, many scientists and media reports say storms are growing in intensity.

This claim, Lomborg argues, also is false.

"[No,] there aren't more powerful hurricanes either. The frequency Category 3 and above hurricanes making landfall since 1900 is also trending slightly down," Lomborg writes. "A July *Nature* paper finds that the increases in strong hurricanes you've heard so much about are 'not part of a century-scale increase, but a recovery from a deep minimum in the 1960s–1980s.'"

Still, not everyone agrees with Lomborg and *Nature*. Some believe that the decline in the number of hurricanes is resulting in hurricanes that indeed are more powerful. But how much more?

Chris Landsea, tropical analysis forecast branch leader at the National Hurricane Center, said global warming likely added about 1 percent more power to Hurricane Michael, a Category 5 hurricane. That translated to 1 or 2 mph.

"That is a fairly small increase and most of the computer guidance by global warming models say maybe we could see 3 percent stronger by the end of the century," said Landsea, speaking during a session on hurricane history in 2019. "That's really not very much."

## Why So Many Crises?

The actual science of global warming and hurricanes seems fairly clear. Hurricanes are not landing more often on the continental US, but less often. It's unclear if they are becoming more powerful, but if hurricanes are growing in intensity, it's not by very much.

These scientific revelations are rather bland, and they seem a stark contrast to claims that extreme weather poses an "existential crisis" to humanity and headlines of "turbocharged" storms.

A person could be forgiven for asking: What gives? What am I supposed to believe? Is a weather apocalypse truly upon us?

If an extreme weather apocalypse is indeed upon us, it is one of many crises we're told we face. There is no shortage of catastrophes and epidemics, judging from politicians, intellectuals, and media reports. Mass shootings. The coronavirus. The opioid crisis. Forest fires. The list goes on.

While it's true conflict and crises are common elements of human history, it seems that our modern state of affairs is virtually constant crises. Why?

In his book *Crisis and Leviathan*, the economist Robert Higgs discusses this phenomenon. Higgs argues that crises are essentially food for the leviathan, a metaphor for the state coined by the English philosopher Thomas Hobbes and derived from a Biblical sea monster.

Higgs observes that crises have served as the catalysts for the greatest expansions of state power in modern history. The New Deal was the spawn of the Great Depression. The War on Terror and the Patriot Act were the offspring of the 9/11 attacks. And then there is the Pandemic of 2020. Unlike in previous pandemics, public health officials leveraged the full power of the state to attempt to tame COVID-19.

In each crisis, Americans were told the emergency actions taken were not just necessary, but temporary. History, however, shows that once a crisis has passed, "the fattened leviathan continues to hold sway."

Higgs's thesis—that crises are the food that feeds the ravenous leviathan, slowly freeing it from the shackles designed to constrain it—calls to mind a meme popular on social media.

"If we let politicians break the law in an emergency," it goes, "politicians will create an emergency so they can break the law."

Another version of the meme would be this: Once crises are seen as a legitimate cause for extraconstitutional action, prepare yourself for an abundance of crises.

## The Best Way to Fight Extreme Weather

None of this is to say pandemics, extreme weather, shootings, and the like are not real or serious problems. They are.

But it's important to understand that government is the cause of many of these problems, not the solution. The reality is government isn't very good at solving simple problems, let alone highly complex ones. Indeed, climate-related deaths are at historic lows—not because governments routinely hit their CO2 reduction benchmarks (they don't) but because free market capitalism has made human habitats exponentially more resistant to climate-related disasters.

"Better infrastructure, fed by improved technology and wealth, does more to protect lives and property than cutting carbon emissions," Lomborg explains.

Global temps may indeed be edging upward, but the solution isn't to give politicians and government bureaucrats the power to regulate the economy with Green New Deal-style legislation designed to curb bad weather.

The solution is to unleash the power of the free market and allow entrepreneurs to build humans a more prosperous and resilient world through human ingenuity.

# Humanities and Science Can Combat the Threat of Climate Change


***Steven D. Allison and Tyrus Miller***

*Steven D. Allison is a professor of ecology and evolutionary biology. Tyrus Miller is Dean, School of Humanities. Both are at the University of California at Irvine.*


Large wildfires in the Arctic and intense heat waves in Europe are just the latest evidence that climate change is becoming the defining event of our time. Unlike other periods that came and went, such as the 1960s or the dot-com boom, an era of unchecked climate change will lead to complex and irreversible changes in Earth's life support systems.

Many people view climate change as a scientific issue—a matter of physical, biological and technical systems. The Intergovernmental Panel on Climate Change's most recent assessment report, for example, is a vast compendium of climate science, threats and potential solutions.

Yet modern climate change is also a human problem caused by the collective behaviors of people—mostly the wealthy—around the world. Japanese economist Yoichi Kaya summarizes this viewpoint in an elegant equation known as the Kaya Identity: Global greenhouse gas emissions are the product not just of energy use and technology, but also human population size and economic activity.

Of course, science is essential for understanding climate change, and technology is critical for solving the problem. But the IPCC report spends little more than 10 pages on climate ethics, social justice and human values. We worry that overemphasis on science may hamper the design of effective climate solutions.

In our view, solving the world's climate problems will require tapping into brainpower beyond science. That's why the two of us—an ecologist and a humanities dean—are teaming up to rethink climate solutions. Recently we developed a program to embed humanities graduate students in science teams, an idea that climate research centers are also exploring.

## A Human-Centered Perspective

Scholars in the humanities interpret human history, literature and imagery to figure out how people make sense of their world. Humanists challenge others to consider what makes a good life, and pose uncomfortable questions—for example, "Good for whom?" and "At whose expense?"

Going beyond science, humanists can define cultural forces driving climate change, such as the fossil fuel dependence of industrialized societies.

In her book, *Living Oil: Petroleum Culture in the American Century*, literature scholar Stephanie LeMenager asserts that 20th-century culture—novels, poetry, films, photography and television—generated a mythology of "petro-utopia." Images of gushing oil derricks implied that the American good life meant unfettered consumption of fossil fuels.

Popular culture, land use and economics reflected this ideal, particularly in California. Even as the Golden State strives to lead the nation in combating climate change, the legacy of petro-culture endures in suburban sprawl and jammed freeways.

Humanist scholars like LeMenager help to uncover the root causes of complex problems. Yes, rising carbon dioxide levels trap more heat in the atmosphere—but values matter too. Defining features of American identity, such as independence, freedom, mobility and self-reliance, have become entangled with petroleum consumption.

## The Softer Side of Technology

When thinking about climate solutions, people often picture technical fixes. The IPCC reports list many ideas for climate change mitigation and adaptation. Mitigation reduces greenhouse gas emissions through technologies like renewable energy. Adaptation, such as building sea walls, aims to manage climate change impacts. It also includes schemes to engineer Earth's climate system—for example, releasing chemicals into the stratosphere to reflect sunlight back into space.

In principle, scientists and engineers could deploy any of these fixes. But should they? To answer this question, society needs humanists and their "soft" technologies—intangible tools for solving problems based on nonscientific knowledge.

Cultural scholars and philosophers can inject ethical principles into policymaking. Relative to emissions reductions, expensive adaptation schemes are less likely to benefit indigenous populations, future generations and the poor—the groups that are most vulnerable to climate change.

Humanists can also help decision makers see how history and culture affect policy options. Plans to improve fuel economy will need to address the historical bond between petroleum and personal freedom. Alternatively, humanity could keep burning fossil fuels while trying to capture the emissions. Yet some societies might balk at the high costs of relatively unproven carbon capture technologies.

## Aligning Climate Solutions with Human Values

So far, scientific facts have not motivated Americans to support the huge societal transformations needed to stop climate change. Some reject the scientific consensus on global warming because it makes them feel bad or clashes with their personal experience of the weather.

Climate change matters more when it affects people's homes, livelihoods and spiritual beliefs. Recent protests against the Dakota Access Pipeline are an example. Opponents condemned the desecration of Native American burial sites and called out violations of historical land treaties dating back 150 years. To them, the pipeline was not just a source of greenhouse gases. It was a threat to their ideals and spirituality.

By tapping into what moves people, the emerging field of environmental humanities can help spur climate action. Scholars of history, philosophy, religious studies, literature and media are exploring many aspects of humans' relationship with the Earth. An entire literary genre of climate fiction, or "Cli-Fi," depicts often-apocalyptic visions of climate impacts on humanity. Social scientists have worked out how civilizations like the ancient Maya and medieval Icelanders dealt with climate shocks.

Together with scientists, environmental humanists are reforming scenarios used in climate modeling. Scenarios originated as an improvisational form of theater, and humanists are reclaiming them as a rehearsal space for the massive societal shifts required to avert dangerous climate change.

## Uniting Humanists and Scientists

We think that stronger collaborations across the humanities and sciences are key for effective climate solutions. Still, there are hurdles to overcome. Humanists have been criticized for failing to apply their expertise to environmental problems outside academic circles. For their part, scientists need to respect humanists as scholars in their own right, not just clever translators of hard science.

In our view, it's time for scientists, engineers and humanists to break down these barriers and appreciate the human element of global climate change.

# It Is Not Too Expensive to Stop Climate Change

*Shahir Masri*

*Shahir Masri, PhD, is an air pollution scientist at the University of California at Irvine.*

As we approach election season, climate change is shaping up to be a prime topic for discussion—and it's no coincidence.

In recent years, we have seen record breaking wildfires, hurricanes, heat waves and floods rip across the country with alarming frequency. Youth climate movements, school walkouts and other demonstrations calling for government action on climate change have sprung up as a result. The 2018 U.S. midterm election even brought with it a team of fresh, young faces who are determined to make climate policy and energy innovation a core issue.

Yet, as climate policy takes the stage, we're already seeing discussions on the economic costs of such policy that make no mention of the glaring elephant in the room; namely, the cost of climate change. Simultaneously, age-old misconceptions pitting the environment against the economy resound.

The so-called Green New Deal, introduced by Rep. Alexandria Ocasio-Cortez (D-N.Y.) and Sen. Ed Markey (D-Mass.) back in February provides a recent example. Much talk about the Green New Deal has centered around massive cost estimates, including an American Action Forum (AAF) study that put a $53 trillion to $93 trillion price tag on the resolution. By picking the upper estimate and rounding up by another $7 trillion, President Trump even produced his own $100 trillion cost estimate. Never mind that it's too premature to estimate the cost of the Green New Deal, given its non-specific language.

"Climate Policy Is Expensive, but So Is Climate Change," by Shahir Masri, The Hill, July 29, 2019. Reprinted by permission.

A more salient point is that these numbers reflect a "cost" analysis, not a "cost-benefit" analysis. That is, the AAF study makes no attempt to weigh the positives of the Green New Deal against the negatives. Put another way, the estimated cost fails to consider the reason we're even talking about the Green New Deal—that is, "climate change." It discusses the upfront economic costs of the resolution without taking into account the costs incurred in the absence of the resolution due to global warming.

Climate-related natural disasters in the U.S. have racked up nearly $800 billion in economic costs in this decade alone, which is up four-fold compared to the 1980s. And natural disasters aren't the only climate-related costs we can expect. The recent U.S. National Climate Assessment estimates future warming to cause up to 2 billion lost labor hours annually by 2090 due to temperature extremes, costing an estimated $160 billion in lost wages.

The report also notes losses in some economic sectors projected to reach hundreds of billions of dollars per year by the end of the century. This doesn't include the enormous public health costs of air pollution from burning fossil fuels, causing around 200,000 deaths in the U.S. annually, or the poisoning of waterways and other instances of environmental destruction that arise from coal mining, fracking, etc. In short, climate change and business-as-usual energy production are expensive.

All told, the annual economic impacts of climate change in the United States will likely cost around 5 percent of GDP by the end of this century, or roughly $1 trillion per year. Over time, these astronomical costs begin to resemble the costs of the Green New Deal. Although, climate mitigation comes with the added perk of preventing catastrophic climate impacts and human mortality. This is not to say the Green New Deal is the perfect way forward. However, it does highlight an imperative piece to be factored in when weighing the costs of climate policy.

Alongside the misleading price tags that tend to accompany proposed environmental policies, there is another age-old industry mantra that deserves correction: that environmental regulation will cripple the economy. History tells a different story. If we look at the last three decades of the 20th century when U.S. environmental law began to seriously grow and take shape, we find that the economy was generally thriving. Harvard Law School Professor Richard Lazarus points out that "the actual cost of pollution controls have . . . almost always proved to be less than those industry projected in initially opposing their imposition."

In 1990, industry estimated that VOC pollution controls applied to stationary sources would cost $15 billion per year. What was the actual cost? Less than $1 billion. In 1989, the utility industry opposed acid rain programs to limit sulfur dioxide emissions, claiming it would cost $4 to $7 billion per year. The actual price tag was about $1 billion to $2 billion per year.

Turning to the automobile industry, strict pollution controls on cars have never been favored, with some manufacturers calling emissions reductions "impossible." Such claims were again exaggerated, as Corporate Average Fuel Economy standards forced innovation that enabled fuel efficiency to increase, tailpipe emissions to decline, and air quality to improve as a result.

And let's not forget the dangerous hole in the ozone layer. When science identified chemicals called CFCs as the culprit, industry quickly moved to their standard defense, claiming that a CFC ban would result in "major economic disruption." Yet, innovation led to a new chemical that was safer for ozone. Our environment was protected, and the economy prospered.

As we now face the even greater challenge of climate change, one must keep this history in mind. We can indeed maintain economic prosperity while protecting our air, waterways, lands, oceans and climate. What's more, we have solutions at our disposal.

While the Green New Deal is but one example, let us not forget that a separate bill exists on the table of Congress today that would dramatically reduce greenhouse gas emissions. The Energy Innovation & Carbon Dividend Act (H.R. 763), if passed, would put at price on atmospheric carbon pollution, and in turn reduce carbon emissions by an estimated 40 percent in the first 12 years.

Such a bill is consistent with key aspects of the Green New Deal and is regarded as more politically viable.

Economists have long agreed that a price on carbon is the most economically efficient way to address climate change. Whether or not Congress will agree will likely depend on public support. Thus, the role of each citizen remains essential if we're to see proposed climate policy ultimately become law.

Current Controversies

Chapter 2

# Is Climate the Only Factor in Changes of Biodiversity?

## Overview: Biodiversity and Its Threats

### *Conserve Energy Future*

*Conserve Energy Future was started to encourage the average person to join in the effort of energy conservation.*

Biodiversity or biological diversity is a term that describes the variety of living beings on earth. In short, it is described as a degree of variation of life. Biological diversity encompasses microorganisms, plants, animals, and ecosystems, such as coral reefs, forests, rainforests, deserts, etc.

Biodiversity also refers to the number or abundance of different species living within a particular region. It represents the wealth of biological resources available to us. It's all about sustaining the natural area made up of a community of plants, animals, and other living things that is being reduced at a steady rate as we plan human activities that are being reduced by habitat destruction.

The United Nations designated 2011–2020 as the United Nations Decade on Biodiversity. In biodiversity, each species, no matter how big or small, has an important role to play in the ecosystem. Various plant and animal species depend on each other for what each offer, and these diverse species ensure natural sustainability for all life forms. Healthy and solid biodiversity can recover from a variety of disasters.

Biodiversity has three essential elements:

- Genetic diversity,
- Ecosystem diversity and
- Species diversity

Recently a new aspect has also been added—"molecular diversity."

Biodiversity is unevenly distributed. It varies globally and within regions. The various factors that influence biodiversity include temperature, altitude, precipitation, soils, and their relation with other species. For instance, ocean biodiversity is 25 times lesser than terrestrial diversity. Biodiversity also increases its form as it moves from the poles towards the tropics.

Biodiversity is the result of 3.5 billion years of evolution. It has been subject to periods of extinction. The latest and most destructive stage of extinction is Holocene extinction, which has occurred due to the impact of human beings on the environment.

## Why Is Biodiversity Important?

The role of biodiversity in the following areas will help make clear the importance of biodiversity in human life:

### *1. Biodiversity Offers a Number of Natural Services*

We humans depend on a number of natural services offered by ecosystems to live a healthy life on Earth. They are as follows.

Maintaining the Balance of the Ecosystem: Recycling and storage of nutrients, combating pollution by breaking it down and its absorption, stabilizing climate, protecting water resources, forming and protecting soil, recovery from unpredictable events and maintaining overall eco-balance.

Provision of Biological Resources: Provision of medicines and pharmaceuticals, food for the human population and animals, ornamental plants, wood products, breeding stock, future resources and diversity of species, ecosystems, and genes.

Social Benefits: Recreation and tourism, cultural value and education, research and monitoring.

### *2. Biodiversity and Food*

Biodiversity provides for a variety of foods for the planet. Because of the availability of different species, humans are able to obtain a range of materials and foods to support their well-being and health.

The various foods such as fish, meat, vegetables, fruits, and cereals are all available because of the planet's biodiversity. 80% of the human food supply comes from 20 kinds of plants. But humans use 40,000 species for food, clothing, and shelter.

### *3. Biodiversity and Human Health*

Biodiversity also plays an essential role in drug discovery and medicinal resources. Through biological diversity, scientists have made significant advances in medical discoveries and have found cures for several diseases.

All this has been possible because of research into the various animal and plant genetics as well as biology. 80% of vaccines and drugs used in prevention and treatment respectively are from the world's biodiversity. Medicines from nature account for usage by 80% of the world's population.

### *4. Keeping Bio-Diverse Ecosystems Intact Helps Humans Stay Healthy*

Research indicates that there is a close link between disease outbreaks and the degradation of nature.

As the global wildlife trade continues and development projects expand deeper into tropical forests, humans are increasing their exposure to wild animals and the diseases they may carry. Seventy percent of emerging viral diseases have spread from animals to humans.

For example, the COVID-19 pandemic can likely be sourced to a wild animal and fish market in Wuhan, China. With COVID-19, we've seen the damage that diseases can do not only to human health but also to the global economy.

Deforestation is also accelerating climate breakdown, which in turn may boost the spread of disease by allowing disease carriers like mosquitoes to extend their geographic ranges and infect new populations of humans.

This shows that we must take care of nature to take care of ourselves. By protecting biodiversity in Earth's ecosystems, countries could save lives and money, while helping to prevent future pandemics.

### *5. Continuity of Planet and Balance of the Ecosystem*

Biodiversity plays an important role in offering ecological services that make life livable on Earth. The ecological balance and biodiversity are crucial for all of the earth, not just humans.

The ecological services include air purification, replenishing and cleaning water systems, absorbing chemicals and breaking down pollutants, stabilizing climate, recycling nutrients and storage, forming and protecting the soil and speedy recovery from natural disasters. Crucial life cycles such as the water cycle and the nitrogen cycle are all determined by biodiversity.

### *6. Biodiversity—A Solution to Climate Change*

Protecting biodiversity plays a crucial part in achieving these emissions reductions and nature can deliver at least 30 percent of the emissions reductions needed by 2030 to prevent climate catastrophe according to a landmark study published in 2017, a group of researchers led by Bronson Griscom, who researches natural climate solutions at Conservation International.

The destruction of forest ecosystems is responsible for 11 percent of all global greenhouse gas emissions caused by humans, so conserving forests would stop the release of these gases into the atmosphere. Trees and plants also store carbon in their tissue, making it even more necessary to protect them.

Some ecosystems, such as mangroves, are particularly good at storing carbon and keeping it out of the atmosphere. Forests and wetland ecosystems provide crucial buffers to extreme storms and flooding related to climate change.

These ecosystems are complex, which means they function best, and are more resilient to the effects of climate change, when all the pieces of the ecosystem are in place, meaning the biodiversity is intact.

### *7. Biodiversity and Industry*

Biological sources provide many industrial materials. The numerous industrial raw materials, including rubber, cotton, leather, food, paper, timber, water, fiber, oil, and dyes, are supplied by the biological resources. These resources are then used by the industries to process and manufacture different products for human and other uses.

### *8. Biodiversity and Economy*

Biodiversity is priceless. However, there have been attempts to put an economic value on biodiversity for people to understand the magnitude of the issue that how important the environment is to humanity and what costs and benefits there can be in doing (or not doing) something.

At least 40 percent of the world's economy and 80 percent of the needs of the poor are derived from biological resources. The Economics of Ecosystems and Biodiversity (TEEB) initiative estimates that global sustainable business opportunities from investing in natural resources could be worth US$2 to 6 trillion by 2050.

Altogether, sectors like pharmaceutical, biotechnology, agricultural seeds, personal care, botanical, food and beverage industries, commercial forestry, and ecotourism industries could lose US$338 billion per year if the loss of biodiversity continues at its current pace.

Around 75 percent of global food crops rely on animals and insects such as bees to pollinate them. Still, many of these pollinator populations are in decline, which could put more than US$235 billion of agricultural products at risk.

Millions of people also depend on nature and species for their day-to-day livelihoods. This is particularly true for struggling communities in developing countries, who often turn to high-biodiversity ecosystems as their source of food, fuel, medicines and other products made from natural materials for their own use and as sources of income.

Nature-related tourism is also a significant income generator for many people as well.

### *9. Biodiversity and Culture*

Biodiversity provides a "wonder" of how things are amazingly inspiring, beautiful, and diverse in nature. Simply because of this, biodiversity promotes recreational activities such as bird watching, fishing, trekking, mountain climbing, and game visits that lead to tourism.

Biodiversity influences cultural values as it inspires people in different ways. Ecosystems such as parks and other protected areas provide recreation and a knowledge resource for visitors, and biodiversity is a frequent source of inspiration for artists, musicians and designers and determines certain lifestyle orientations. Biological education and research are as a result of the existing biodiversity.

Species are frequently integral to religious, cultural and national identities. All major religions include elements of nature, and 231 species are formally used as national symbols in 142 countries.

Unfortunately, more than one-third of those species are threatened, but the bald eagle and American bison are examples of conservation successes because of their role as national symbols.

### *10. Adjustment and Adaptation*

The biodiversity in genetic makeup helps plants and animals to adapt and adjust to respective environmental changes. Genetic diversity, for instance, helps species to fight diseases.

## Reason for Loss of Biodiversity

The earth's biodiversity is in grave danger. In the present era, human beings are the most dangerous cause of the destruction of the earth's biodiversity. In 2006, the terms threatened, endangered, or rare were used to describe the status of many species.

The "evil quartet" identified by Jared Diamond is overkill, habitat destruction, secondary extinctions, and introduced species. Factors identified by Edward Wilson are described by the acronym-HIPPO standing for habitat destruction, climate change, invasive species, pollution, human overpopulation, and over-harvesting.

### 1. Habitat Destruction

Habitat destruction is a major cause of biodiversity loss. Habitat loss is caused by deforestation, overpopulation, pollution, and global warming. Species that are physically large and those living in forests or oceans are more affected by habitat reduction.

Some experts estimate that around 30% of all species on earth will be extinct by 2050. According to the International Union for Conservation of Nature (IUCN), globally, about one-third of all known species are threatened with extinction. Even it is estimated that 25% of all mammals will be extinct within 20 years.

### 2. Invasive Species

Even if a small element of an ecosystem breaks down, the whole system's balance is threatened. Freshwater ecosystems are nowadays the most threatened ecosystems. Invasive species refer to those that would normally remain constrained from an ecosystem because of the presence of natural barriers.

Since these barriers are no longer existing, invasive species invade the ecosystem, destroying native species. Human activities have been the major cause of encouraging invasive species.

### 3. Over-Exploitation of Species

Species can also be threatened by genetic pollution—uncontrolled hybridization and gene swamping. For instance, abundant species can interbreed with rare species, thus causing swamping of the gene pool.

Overexploitation is caused by activities such as overfishing, overhunting, excessive logging and illegal trade of wildlife. Over 25% of global fisheries are being overfished at unsustainable levels.

### 4. Global Warming and Climate Change

Global warming and climate change are also becoming a major cause of loss of biodiversity. Changes in climates and global temperatures directly impact physical environmental factors essential for sustainable habitat. For example, if the present rate of global warming continues, coral reefs, which are biodiversity hotspots, will disappear in 20-40 years.

The wildlife in the mountain regions that requires cool temperatures of high elevations such as the rock rabbit and mountain gorillas may in the near future run out of habitat due to global warming. If global warming and climate change continue, 10% of the entire world species might go extinct by 2050.

### *5. Pollution*

The various forms of pollution, including water pollution, soil pollution, air pollution, land pollution and agricultural pollution, pose a serious threat to the biological systems destroying animal and plant habitats due to the release of toxic substances and chemicals.

Some seriously polluted regions have become dead zones as the conditions cannot sustain any life form. Apart from habitat destruction, pollution poses long-term cumulative impacts on the species' health, contributing to their eventual death. For instance, marine and freshwater life forms are most affected by pollution.

### *6. Human Overpopulation*

Overpopulation has witnessed continued encroachment into frontier forests, heightened pollution, and destruction of natural ecosystems that have considerably contributed to the mass extinction of species. The number of threatened species persists in multiplying worldwide, whereas some have completely gone extinct.

Human activities such as acidifying water systems, over-exploitation of natural resources, pollution, over-fishing, poaching, and the deliberate and indirect destruction of natural systems have contributed to the loss of biodiversity.

### *7. Natural Calamities*

Natural calamities, such as floods, drought, forest fires, earthquakes, volcanic eruptions, epidemics etc. sometimes take a heavy toll on plant and animal life. In nature, such episodes are usually confined to specific plant or animal populations as the pathogen is often specific to particular species or group of species.

Floods are frequent in moist tropical regions, which tend to inundate much of the ground vegetation, trap a large number of animals while leading away soil nutrients. In densely wooded localities, forest fires often reduce a large number of plant and animal species to ashes, and so do earthquakes.

Volcanic eruptions may, at times completely destroy plant and animal life in its surrounding areas. Epidemics sometimes destroy large portions of a natural population.

### *8. Genetic Pollution*

Genetic pollution refers to the hybridization or genetic engineering of species. It is applied mainly in agricultural production to increase resistance to diseases and local climates for high yields. This threatens species, especially when there are uncontrolled hybridization and genetic engineering. Eventually, it brings about unique genotypes which replace the originally existent diverse genetic materials.

Thus we can see that biodiversity, which is crucial for the well-being of life on earth, is coming under the threat of many factors related to human activities. There is an urgent need to take action to protect the magnificent biodiversity of our planet. We must create economic policies to maintain the Earth's biodiversity and take appropriate measures to protect habitats and species.

# Climate Change Affects the Future of Life on the Planet

*Alex Hubbe and Mark Hubbe*

*Alex Hubbe is a paleontologist and an evolutionary biologist and associate professor at the Instituto de Geociências, Universidade Federal da Bahia, Brazil. Mark Hubbe is associate professor in the Department of Anthropology at Ohio State University.*

Do you know the difference between weather and climate? The weather is composed of a series of physical properties, like temperature, rainfall, wind speed, and many other characteristics that change according to seasons, days, or even hours. When we consider the weather over long periods of time (30 years or more) we are talking about climate. Our planet's climate is constantly changing over time. Some time periods in the past were considerably warmer and others considerably colder than today and similar changes will happen in the future. Around 20,000 years ago, the last cooling phase ended, starting a warming period until around 8,000 years ago, during which Earth's average temperature rose between 5.6 and 8.5°F (3.1–4.7°C).[1] This means a change of around 0.06°F (0.03°C) every century. Between 8,000 years ago and the Industrial Revolution (around 200 years ago), temperatures were relatively constant. Since then, temperatures have been increasing again, and very quickly. By 2100, scientists believe that the mean temperature on the planet will have raised an additional 3.6–8.8°F (2.0–4.9°C) compared with the temperatures at the time of the Industrial Revolution.[2] This means that the planet's temperature rise in the next century will be about 100 times faster than what happened during the last 20,000 years. While past changes in climate were the result of several factors, like the amount of

Originally published as: "Current Climate Change and the Future of Life on the Planet," by Alex Hubbe and Mark Hubbe, Frontiers Media S.A, 03/07/2019. https://kids.frontiersin.org/articles/10.3389/frym.2019.00037. 

energy coming from the sun and the concentration of different atmospheric gases, this time humans are mainly responsible. And that is why we call the current warming the anthropogenic (or human-induced) climate change.

Since the Industrial Revolution, our species has been burning immense amounts of fossil fuel to support our ever-growing demand for energy. Primarily through this activity, over the last 200 years, humans have increased the amount of carbon dioxide (CO2) in the atmosphere by about 46%. And CO2 is not the only factor affecting climate. The use of fossil fuels, agriculture, industrial processes, and deforestation emit other substances (or prevent them from naturally decreasing) that contribute to climate changes.

The anthropogenic climate change may not seem much to us now, but it will have a huge impact on humankind within the current and next few generations. The frequency and intensity of natural phenomena, like droughts and storms, will very likely increase with several consequences to humans. Even our shorelines will dramatically change as the oceans rise, primarily due to the melting of the ice caps. This may result in the abandonment of entire coastal cities, like New York, Miami, Rio de Janeiro, Amsterdam, and Bangkok.

Climate change will also affect life forms on the planet. Several microbes, plants, and animals will probably become extinct. Just to give you a glimpse of how life can change in response to climate change, only 6,000 years ago the Sahara Desert, one of the driest places on Earth today, was covered by lush vegetation. This green landscape was a result of the more humid conditions back then.[3]

## So, What Happens to Living Beings When the Climate Changes?

Climate changes affect how living beings interact with each other and with the environment they live in. When faced with climate challenges, species have two main options (which can also happen in combination). The first is to migrate. For example, as the ocean surface temperatures increase due to the anthropogenic climate change, several species, from algae to fishes, are moving toward the poles in search of colder waters.[4] For migration to happen, species must be able to find another place where they can survive and prosper. This is easier, as you can imagine, for species that are able to travel long distances during their lifetimes, like whales and cougars. However, it can be very hard for species that cannot travel long distances or that do not disperse seeds very far or very often, like sloths and orchids.

The other way a species can survive is to evolve. Evolution can happen on a small scale. For example, over a few generations, alpine chipmunks (*Tamias alpinus*) at Yosemite National Park, California, USA, evolved longer faces over 100 years in response to climate change.[5] Evolution can also happen on a large scale, resulting in the origin of new species. For instance, brown bears (*Ursus arctos*) and polar bears (*Ursus maritimus*) evolved from the same species of bear in a process that started around 480–340 thousand years ago, as the two groups accumulated more and more differences over time.[6]

Sadly, when species cannot migrate or evolve in response to the changes in the environment, they become extinct. Human lifestyle and the rapidly increasing number of humans on the planet impose a series of challenges for the survival of several species. Around two vertebrate species became extinct per year during the last 100 years, mainly due to human causes.[7] Some examples of the many harmful byproducts of our lifestyle are the intense deforestation due to agriculture and the pollution of air and water. Anthropogenic climate change deserves special attention, because it is different from other human impacts. Anthropogenic climate change has a global effect, even on areas that are seldom disturbed by other human impacts, such as some protected areas. In addition, anthropogenic climate change is happening too quickly for several species to be able to migrate or evolve. This means that anthropogenic climate change is a serious threat to species that are already struggling to survive on an extensively human-transformed Earth. The take-home message here is: humans are causing a long-lasting and devastating impact on life on Earth. To put it in numbers, it is possible that over 75% of all species, three out of every four species that currently exist, will be extinct a few centuries from now![8]

Extinction of exquisite life forms is sad in itself, but the problem of extinction goes far beyond just the disappearance of these organisms. We rely on many of them to survive. The quality of the air we breathe, the water we drink, and the food we eat, just to name a few, are directly related to the existence and the health of several other species.

In addition, we also study other lifeforms to find solutions to our own problems. For instance, the hook-and-loop fastener (a.k.a. VELCRO) was developed after studying the burrs (seeds) of a plant called burdock, which naturally attach to clothes. Several medicines have been discovered in different life forms and were only later synthesized in laboratories. There are countless inventions/discoveries waiting for us to find by studying nature, but they are only available while the species that can help us are still alive on the planet.

## Be Part of the Solution!

All over the world, people are working together to lessen the effects of climate change and other negative human impacts on the planet. The Paris Agreement, which started in 2016, is a good example of this. This agreement is a global effort to prevent global temperatures from rising more than 3.6°F (2°C) in relation to the period before the Industrial Revolution. The basic idea is to drastically reduce global emissions of CO2 and other substances that contribute to climate change. Currently, 185 countries out of almost 200 have joined in support of the Paris Agreement. The Danish city of Copenhagen has an ambitious plan, too. It wants to become a carbon-neutral city by 2025. To this end, among other things, the city is changing its energy production from fossil fuel to mainly wind- and animal/plant waste-derived energy, reducing the amount of trash produced, and transforming the city in a huge bicycle path. In several countries, scientists have been studying different ways to remove CO2 from the atmosphere and turn it into something useful. Some researchers, for example, are studying how to use bacteria to turn CO2 into fuel.

However, in a planet with over 7.6 billion people, we all need to change our ways of living to minimize anthropogenic climate change. There are many ways that each of us can do our part. A good first step is to calculate your family's carbon footprint. The carbon footprint measures how much CO2 we are emitting into the atmosphere over a certain period of time. You can calculate your carbon footprint using different calculators online. One of the many possibilities is https://coolclimate.berkeley.edu/calculator. And once you know your carbon footprint, try to reduce it! You will be amazed by how many things we do that impact the planet. The bright side is that there are also many simple ways to reduce our negative impact. Here are some easy things any of us can do to help prevent anthropogenic climate change: http://www.un.org/sustainabledevelopment/takeaction/. A few of these easy actions include: do not buy things you do not need, unplug appliances and turn off the lights you are not using, take shorter showers, eat less meat, poultry, and fish, shop locally and from sustainable sources, and use your bike or public transportation instead of a car whenever possible. Simple actions like these can make a big difference if enough of us perform them! We should all try to reduce our carbon footprints as much as we can, because we will then reduce our impact on anthropogenic climate change and the related negative consequences for life on Earth.

Let us change the world in a cooler way!

## *References*

1. Annan, J. D., and Hargreaves, J. C. 2013. A new global reconstruction of temperature changes at the last glacial maximum. *Clim. Past* 9:367–76. doi: 10.5194/cp-9-367-2013

2. Raftery, A. E., Zimmer, A., Frierson, D. M. W., Startz, R., and Liu, P. 2017. Less than 2°C warming by 2100 unlikely. *Nat. Clim. Change* 7:637–41. doi: 10.1038/nclimate3352

3. deMenocal, P., Ortiz, J., Guilderson, T., Adkins, J., Sarnthein, M., Baker, L., et al. 2000. Abrupt onset and termination of the African humid period: rapid climate responses to gradual insolation forcing. *Q. Sci. Rev.* 19:347–61. doi: 10.1016/S0277-3791(99)00081-5

4. Poloczanska, E. S., Brown, C. J., Sydeman, W. J., Kiessling, W., Schoeman, D. S., Moore, P. J., et al. 2013. Global imprint of climate change on marine life. *Nat. Clim. Change* 3:919–25. doi: 10.1038/nclimate1958

5. Walsh, R. E., Assis, A. P. A., Patton, J. L, Marroig, G., Dawson, T. E., and Lacey, E. A. 2016. Morphological and dietary responses of chipmunks to a century of climate change. *Glob. Change Biol.* 22:3233–52. doi: 10.1111/gcb.13216

6. Liu, S., Lorenzen, E. D., Fumagalli, M., Li, B., Harris, K., Xiong, Z., et al. 2014. Population genomics reveal recent speciation and rapid evolutionary adaptation in polar bears. *Cell* 157:785–94. doi: 10.1016/j.cell.2014.03.054

7. Ceballos, G., Ehrlich, P. R., and Dirzo, R. 2017. Biological annihilation via the ongoing sixth mass extinction signaled by vertebrate population losses and declines. *Proc. Natl. Acad. Sci. U.S.A.* 114:E6089–96. doi: 10.1073/pnas.1704949114

8. Barnosky, A. D., Matzke, N., Tomiya, S., Wogan, G. O. U., Swartz, B., Quental, T. B., et al. 2011. Has the earth's sixth mass extinction already arrived? *Nature* 471:51–7. doi: 10.1038/nature09678

# Changes in Weather and Climate Negatively Affect the World's Oceans

*Anup Shah*

*Anup Shah is a computer scientist and the founder of Global Issues, a website dedicated to raising awareness of social, political, and environmental issues that impact the world.*

The link between climate change and biodiversity has long been established. Although throughout Earth's history the climate has always changed with ecosystems and species coming and going, rapid climate change affects ecosystems and species ability to adapt and so biodiversity loss increases.

From a human perspective, the rapid climate change and accelerating biodiversity loss risks human security (e.g. a major change in the food chain upon which we depend, water sources may change, recede or disappear, medicines and other resources we rely on may be harder to obtain as the plants and flora they are derived from may reduce or disappear, etc.).

The UN's Global Biodiversity Outlook 3, in May 2010, summarized some concerns that climate change will have on ecosystems:

> Climate change is already having an impact on biodiversity, and is projected to become a progressively more significant threat in the coming decades. Loss of Arctic sea ice threatens biodiversity across an entire biome and beyond. The related pressure of ocean acidification, resulting from higher concentrations of carbon dioxide in the atmosphere, is also already being observed.
>
> Ecosystems are already showing negative impacts under current levels of climate change … which is modest compared to future projected changes…. In addition to warming temperatures, more frequent extreme weather events and changing patterns of

"Climate Change Affects Biodiversity," by Anup Shah, Global Issues, January 19, 2014. Reprinted by permission.

> rainfall and drought can be expected to have significant impacts on biodiversity.
>
> —Secretariat of the Convention on Biological Diversity (2010), Global Biodiversity Outlook 3, May 2010, p. 56

Some species may benefit from climate change (including, from a human perspective, an increases in diseases and pests) but the rapid nature of the change suggests that most species will not find it as beneficial as most will not be able to adapt.

## Climate Change Impacts on Biodiversity in the Arctic

The Arctic, Antarctic and high latitudes have had the highest rates of warming, and this trend is projected to continue, as the above-mentioned Global Biodiversity Outlook 3 notes (p. 56).

In the Arctic, it is not just a reduction in the extent of sea ice, but its thickness and age. Less ice means less reflective surface meaning more rapid melting. The rapid reduction exceeds even scientific forecasts and is discussed further on this site's climate change introduction.

In terms of biodiversity, the prospect of ice-free summers in the Arctic Ocean implies the loss of an entire biome, the Global Biodiversity Outlook notes (p. 57).

In addition, whole species assemblages are adapted to life on top of or under ice—from the algae that grow on the underside of multi-year ice, forming up to 25% of the Arctic Ocean's primary production, to the invertebrates, birds, fish and marine mammals further up the food chain. The iconic polar bear at the top of that food chain is therefore not the only species at risk even though it may get more media attention.

Note, the ice in the Arctic does thaw and refreeze each year, but it is that pattern that has changed a lot in recent years.

It is also important to note that loss of sea ice has implications on biodiversity beyond the Arctic, as the Global Biodiversity Outlook report also summarizes:

> Bright white ice reflects sunlight.
>
> When it is replaced by darker water, the ocean and the air heat much faster, a feedback that accelerates ice melt and heating of surface air inland, with resultant loss of tundra.
>
> Less sea ice leads to changes in seawater temperature and salinity, leading to changes in primary productivity and species composition of plankton and fish, as well as large-scale changes in ocean circulation, affecting biodiversity well beyond the Arctic.
>
> —Secretariat of the Convention on Biological Diversity (2010), Global Biodiversity Outlook 3, May 2010, p. 57

## Climate Change Means Ocean Change

When talking about the impacts of climate change, we mostly hear about changes to land and the planet's surface or atmosphere. However, most of the warming is going into the oceans where a lot of ecosystem changes are also occurring:

- 93% Oceans
- 2.3% Atmosphere
- 2.1% Continents
- 0.9% Glaciers and Ice Caps
- 0.8% Arctic Sea Ice
- 0.2% Greenland Ice Sheet
- 0.2% Antarctic Ice Sheet

Indeed, the warming in the oceans has been occurring for quite some time. The implications of this is further explained with Inter Press Service's freezer analogy: The world's northern freezer is on rapid defrost as large volumes of warm water are pouring into the Arctic Ocean, speeding the melt of sea ice.

One of John Bruno's colleagues, Ove Hoegh-Guldberg, talks about the impact climate change will have on ocean ecosystems. A summary says that

> Rapidly rising greenhouse gas concentrations are driving ocean systems toward conditions not seen for millions of years, with

an associated risk of fundamental and irreversible ecological transformation. Changes in biological function in the ocean caused by anthropogenic climate change go far beyond death, extinctions and habitat loss: fundamental processes are being altered, community assemblages are being reorganized and ecological surprises are likely.

## Increasing Ocean Acidification

Although it has gained less mainstream media attention, the effects of increasing greenhouse emissions—in particular carbon dioxide—on the oceans may well be significant.

As explained by the US agency, the National Oceanic and Atmospheric Administration (NOAA), the basic chemistry of ocean acidification is well understood.

These are the 3 main concepts:

1. More CO2 in the atmosphere means more CO2 in the ocean;
2. Atmospheric CO2 is dissolved in the ocean, which becomes more acidic; and
3. The resulting changes in the chemistry of the oceans disrupts the ability of plants and animals in the sea to make shells and skeletons of calcium carbonate, while dissolving shells already formed.

Scientists have found that oceans are able to absorb some of the excess CO2 released by human activity. This has helped keep the planet cooler than it otherwise could have been had these gases remained in the atmosphere.

However, the additional excess CO2 being absorbed is also resulting in the acidification of the oceans: When CO2 reacts with water it produces a weak acid called carbonic acid, changing the sea water chemistry. As the Global Biodiversity Outlook report explains, the water is some 30% more acidic than pre-industrial times, depleting carbonate ions—the building blocks for many marine organisms.

In addition, concentrations of carbonate ions are now lower than at any time during the last 800,000 years. The impacts on ocean biological diversity and ecosystem functioning will likely be severe, though the precise timing and distribution of these impacts are uncertain. (See p. 58 of the report.)

Although millions of years ago CO2 levels were higher, today's change is occurring rapidly, giving many marine organisms too little time to adapt. Some marine creatures are growing thinner shells or skeletons, for example. Some of these creatures play a crucial role in the food chain, and in ecosystem biodiversity.

Some species may benefit from the extra carbon dioxide, and a few years ago scientists and organizations, such as the European Project on Ocean Acidification, formed to try to understand and assess the impacts further.

One example of recent findings is a tiny sand grain-sized plankton responsible for the sequestration of 25–50% of the carbon the oceans absorb is affected by increasing ocean acidification. This tiny plankton plays a major role in keeping atmospheric carbon dioxide (CO2) concentrations at much lower levels than they would be otherwise so large effects on them could be quite serious.

Other related problems reported by the Inter Press Service include more oceanic dead zones (areas where there is too little oxygen in the sea to support life) and the decline of important coastal plants and forests, such as mangrove forests that play an important role in carbon absorption. This is on top of the already declining ocean biodiversity that has been happening for a few decades now.

Scientists now believe that ocean acidification is unparalleled in the last 300 million years, raising the possibility that we are entering an unknown territory of marine ecosystem change.

## Increasing Ocean Stratification

As climate change warms the oceans (even just an increase of about 0.2C per decade, on average), the warmer water (which is lighter) tends to stay on top of what is then a layer of colder water.

This affects tiny drifting marine organisms known as phytoplankton. Though small, phytoplankton are a critical part of our planetary life support system. They produce half of the oxygen we breathe, draw down surface CO2, and ultimately support all of our fisheries, says Boris Worm of Canada's Dalhousie University and one of the world's leading experts on the global oceans (quoted by Inter Press Service—IPS.)

In the same news report, IPS explains that phytoplankton can only live in the top 100 or 200 meters of water, but if it is getting warmer, they eventually run out of nutrients to feed on unless the cold, deeper waters mix with those near the surface.

Ocean stratification has been widely observed in the past decade and is occurring in more and larger areas of the world's oceans, IPS also adds.

Researchers have found a direct correlation between rising sea surface temperatures and the decline in phytoplankton growth around the world.

As NASA summarizes, stratification cuts down the amount of carbon the ocean can take up.

## Increasing Oceanic Dead Zones

The past half-century has seen an explosive growth in aquatic dead zones, areas too low in dissolved oxygen to support life.

Fertilizer and sewage run-off cause huge growth of plankton. However, these then quickly die and are consumed by bacteria that deplete waters of oxygen. For example, the Gulf of Mexico has a 22,000 square kilometer dead zone every spring due to run-off from the Mississippi River.

There is also a linkage with climate change:

> Ocean stratification, where warm water sits firmly on top of cold, nutrient-rich water, also creates dead zones and lowers the overall productivity of the oceans.… Such dead zones were rare 40 years ago but now number several hundred. Without urgent action, climate change will continue to warm oceans, increasing stratification and producing larger and more dead

zones with a major impact on future fisheries, a 2009 study in Nature Geoscience warned.

It will take a thousand years for the oceans to cool down, so it is imperative to pull the emergency brake on global warming emissions, the study concluded.

—Stephen Leahy, Ocean Losing Its Green, Inter Press Service, July 31, 2010

## Coral Reefs Threatened by Climate Change

Around the world, coral reefs have been dying largely due to climate change.

At the beginning of September 2009, the Australian agency looking after the Great Barrier Reef released an outlook report warning the Great Barrier Reef is in trouble.

But it is not just the Great Barrier Reef at risk. All of them are at risk, says Charlie Veron, an Australian marine biologist who is widely regarded as the world's foremost expert on coral reefs.

The future is horrific, he says. There is no hope of reefs surviving to even mid-century in any form that we now recognize. If, and when, they go, they will take with them about one-third of the world's marine biodiversity. Then there is a domino effect, as reefs fail so will other ecosystems. This is the path of a mass extinction event, when most life, especially tropical marine life, goes extinct.

## Lizards Threatened by Climate Change

What the BBC described as a global-scale study published in the journal *Science* found that climate change could wipe out 20% of the world's lizard species by 2080.

Global projection models used by the scientists suggested that lizards have already crossed a threshold for extinctions caused by climate change.

The fear of lowland species moving to higher elevations has long been predicted as an effect of climate change. This has been observed with lizard populations too, as the leader of the research team told the BBC: "We are actually seeing lowland species moving upward in elevation, slowly driving upland species extinct, and if the upland species can't evolve fast enough then they're going to continue to go extinct."

Why are lizards so sensitive to climate change? The BBC summarizes:

> Lizards, the researchers say, are far more susceptible to climate-warming extinction than previously thought. Many species live right at the edge of their thermal limits.
>
> Rising temperatures, they explained, leave lizards unable to spend sufficient time foraging for food, as they have to rest and regulate their body temperature.
>
> —Victoria Gill, Climate change link to lizard extinction, BBC, May 14, 2010

## Other Examples

The above areas of biodiversity affected is by no means exhaustive. Other areas affected by climate change include terrestrial animals and forests, water sources and related ecologies, and so on.

# Abnormal Weather Causes Problems for Bees, Birds, and Bats

*Kylie Johnson*

*Kylie Johnson is the editor at Green and Growing. She contributed this article as a guest writer.*

Climate change has had a profound effect on the Earth's weather. Climate change has impacted the natural rhythms of weather, the behavior of animals, and the blooming of flowers. One of the most impacted populations by climate change are pollinators who are facing unprecedented difficulties as they struggle to survive.

## First of All, What Is Climate Change?

Climate is the long-term weather averages, as opposed to weather, which is the short-term observances of the local climate over hours and weeks. Climate is far more difficult for people to understand as climate data is tracked over years, decades, and centuries. Therefore, climate change is observed drastic changes from the norm in the world's weather patterns during a long period of time. Scientists are able to compare one area's summer to a summer decades before in the same area and are able to determine whether it was drier or wetter than average. Climate change has more drastic effects that can be immediately observed; for example, hurricanes like Hurricane Harvey that occurred in late summer 2017 was fueled by warmer waters and, consequently, was more destructive than hurricanes in the past.

Humans all over the world have been drastically impacted by climate change as the weather during the seasons has become more extreme than it has been in the past. In some areas, summer has become drier and hotter, and in other areas winter is lasting

"How Climate Change Has Affected Pollinators," by Kylie Johnson, Green and Growing, for the Endangered Species Coalition, May 16, 2018. Reprinted by permission.

longer with far more storms and blizzards. Humans have struggled to adapt to these changes as the violent unpredictable weather has become more common than it has in the past.

If a more adaptable and resilient species like human beings are having difficulty adapting to climate change, it becomes far more understandable why a more fragile group such as pollinators are having immense difficulty coping and surviving.

## Impact on Bee Populations

Bees, in particular, are facing the danger of a phenomenon called colony collapse disorder (CCD). A paper from Oregon State University explains CCD: "CCD most likely stems from a combination of problems associated with agricultural beekeeping, including pathogens, nutritional deficiencies and lack of a varied diet, exposure to neonicotinoid insecticides and other pesticides, lack of genetic diversity, habitat loss, and transportation stress. Pesticides, stress, and lack of diversity can actually exacerbate the vulnerability of bees to pathogens." Habitat loss, nutritional deficiencies, and lack of varied diet are tied directly to climate change as the abnormal climate is impacting the growth of plants and flowers. Climate change is making flowers bloom half a day earlier each year, which means that plants are now blooming a month earlier than 45 years ago. Plants blooming earlier ultimately means that they do not get pollinated and bees are left without food.

As bee colonies continue to be affected by CCD, humans will continue to see the impact of CCD on their plants and crops. The website GreenLivingIdeas states, "NEEF reports that about 1,000 plants we depend on for food and products need to be pollinated by animals, including coffee, tasty snacks like melon, chocolate and almonds, and even tequila." According to the website, 75% plants in our yards depend on insects and animals to pollinate them, and bees in particular "add more than $15 billion in value to US agricultural crops each year through pollination." If this dire trend of CCD continues, humans will face severe food shortages and grave economic consequences, not to mention the tragic ramification of bees becoming extinct.

## Impact on Hummingbird Populations

Other pollinators besides bee populations are being affected by climate change as well. Four species of hummingbirds in North America are at risk because of the rising temperatures: Allen's Hummingbird, Black-chinned Hummingbird, Calliope Hummingbird, and Rufous Hummingbird. The increasing warmer temperature is forcing these four species to abandon their native areas for more cooler and stable environments. Intense heat is incredibly dangerous for hummingbirds as it forces them to find shade to cool off rather than feed on nectar, which consequently means that they could starve since their high metabolism demands that they constantly need to eat.

## Impact on Bat Populations

Bats are yet another species of pollinators that are affected by the changing climate. The warmer weather impacts their hibernation cycles and their prey availability, which directly affect how successfully a mother bat can give birth and raise her young. According to *National Geographic*, climate change is also impacting their ultrasonic hearing: "bats living in temperate zones were more likely to lose prey detection volume, while in tropic zones, many bat species will actually be able to detect more prey. Bats calling at lower pitches generally gained prey detection space" because humidity and temperature directly impact how effectively bats can detect their prey.

## What You Can Do to Help Pollinators

- Plant a variety of pollinator friendly flowers and plants that are native to your climate.
- Stop or limit the use of pesticides on your property—pesticides are toxic to pollinators.

- Create a habitat that is friendly to bees. This means either placing beehives on your property, leaving dead logs around that bees can nest in, and simply ensuring bees have plenty of bee-friendly plants to feed from in your yard.
- Providing nectar for hummingbirds on your property. You can do this by buying a feeder for hummingbirds and filling it with sugar water.
- Placing a bat house on your property. This will provide bats a safe place to sleep during the day.
- Plant milkweed plants—Monarch butterflies lay their eggs on the leaves and feed on the nectar of the flowers.

# Deforestation Increases Climate Imbalance and Biodiversity Loss

### *Conserve Energy Future*

*Conserve Energy Future was started to encourage the average person to join in the effort of energy conservation.*

With the world growing at a pace hard to match, the increasing need for space is turning out to be an area of concern. With a desperate need for land for agricultural, industrial and most importantly, urban requirements to contain cities and their growing population, a direct action that we have come to recognize as "Deforestation" occurs.

Deforestation, in simple terms, means the felling and clearing of forest cover or tree plantations to accommodate agricultural, industrial or urban use. It involves the permanent end of forest cover to make that land available for residential, commercial or industrial purposes.

According to Wikipedia:

> Deforestation, clearance, clearcutting or clearing is the removal of a forest or stand of trees from land which is then converted to a non-forest use. Deforestation can involve conversion of forest land to farms, ranches, or urban use. The most concentrated deforestation occurs in tropical rainforests.

Over the last century, the forest cover around the globe has been significantly compromised, leaving the green cover down to an all-time low of about 30 percent. According to the United Nations Food and Agriculture Organization (FAO), an estimated 18 million acres (7.3 million hectares) of forest are lost each year.

Deforestation can also be seen as removal of forests leading to several imbalances, both ecologically and environmentally. What makes deforestation alarming is the immediate and long term

"Deforestation: Compromises of a Growing World," Conserve Energy Future. Reprinted by permission.

effects it is bound to inflict if continued at the current pace. Some predictions state that the rainforests of the world will be eradicated if deforestation continues at its current pace.

Deforestation or clearance occurs due to several reasons. To get a clear overview, we could include the need for money, both in terms of profitability as well as providing for one's family in most scenarios, along with lack of or no forest laws, need for land space for housing etc. among a long list of other uses.

Although primarily blamed on agricultural or pastoral use, farmers usually cut trees for increasing space for cultivation and as fodder land for grazing and surviving livestock. The whole concept of "slash and burn" agriculture is used to indicate this same process where farmers employ the above chain of actions for their purposes.

## Primary Causes of Deforestation

### *1. Agricultural Activities*

As earlier mentioned in the overview, agricultural activities are one of the significant factors affecting deforestation. According to the FAO, agriculture leads to around 80% of deforestation.

Due to the overgrowing demand for food products, a huge amount of trees are felled to grow crops, and 33% of agriculture-caused deforestation is because of subsistence agriculture.

### *2. Livestock Ranching*

Livestock is believed to be responsible for about 14% of global deforestation. Farmers often clear the land by cutting down trees and burning them to raise livestock and grow food. They continue to use the property until the soil is completely degraded and repeat the same process on new woodland.

Eventually, it'll reforest, but it will take many years to return to its original condition. Surprisingly, over the past 40 years, the forest area has reduced by almost 40 percent, and during the same period, pasture regions and cattle populations have grown significantly and rapidly.

### *3. Illegal Logging*

Apart from this, wood-based industries like paper, match-sticks, furniture, etc. also need a substantial amount of wood supply. Wood is used as fuel, both directly and indirectly.

Therefore trees are chopped for meeting the demand for supplies. Firewood and charcoal are examples of wood being used as fuel. Some of these industries thrive on illegal wood cutting and felling of trees.

### *4. Urbanization*

Further, to gain access to these forests, the construction of roads is undertaken; here again, trees are chopped to build roads. Overpopulation too directly affects forest covers, as with the expansion of cities, more land is needed to establish housing and settlements. Therefore forest land is reclaimed.

### *5. Desertification of Land*

Some of the other factors that lead to deforestation are also partly natural and partly anthropogenic, like desertification of land. It occurs due to land abuse, making it unfit for the growth of trees. Many industries in petrochemicals release their waste into rivers, which results in soil erosion and make it unfit to grow plants and trees.

### *6. Mining*

Oil and coal mining requires a considerable amount of forest land. Apart from this, roads and highways have to be built to make way for trucks and other equipment. The waste that comes out from mining pollutes the environment and affects the nearby species.

### *7. Forest Fires*

Another valid example would be forest blazes; hundreds of trees are lost each year due to forest fires in various portions of the world. It happens due to extreme warm summers and milder winters. Fires, whether caused by man or nature, result in a massive loss of forest cover.

### *8. Paper*

According to the Environment Paper Network, the paper that's thrown away each year accounts for approximately 640 million trees. America, China, Canada, Japan, constitute more than that of the world's paper production, and that is 400 million tons a year.

If we recycled, that could save 27.5 million tons of carbon dioxide from going into the atmosphere. We allow the forests to continue to remain as a favorable ecosystem and wildlife habitat if we use recycled paper.

### *9. Overpopulation*

The overpopulation requires more land to establish housing and settlements. It generates a significant need for food and farmland to grow food and raise livestock. It automatically requires many more roads and highways for transport and communication—all these results in deforestation. Logging industries cut down trees for furniture, paper, building materials, and many more products.

Moreover, the growing human population is directly linked to deforestation. Therefore, it becomes almost essential to purchase from sustainable companies that actively work against deforestation.

## Effects of Deforestation

### *1. Climate Imbalance and Climate Change*

Deforestation also affects the climate in many ways. Forests are the lungs of our planet. Trees take in carbon dioxide and release oxygen and water vapor in the air, and that is why tropical rainforests are extremely humid.

Trees also provide shade that keeps the soil moist. All these are compromised with the lack of trees. It leads to the imbalance in the atmospheric temperature, drier climate, further making conditions for the ecology difficult that leads to climate change.

Several animals and plant species that form the flora and fauna across the world are vastly accustomed to their natural habitat. Therefore, haphazard clearance of forests would make it very difficult for them to survive or to shift from their native environment or adapt to new habitats.

When a forest is cut down, the humidity levels come down and cause the remaining plants to dry out. The drying out tropical rainforests increases fire damage that destroys forests rapidly and harms wild animals as well as humans.

Forests and climate are linked intrinsically. Forest loss and degradation are both a cause and an effect of our changing climate. At the same time, deforestation is self-perpetuating.

Therefore, these occurrences are dangerous and fuel further deforestation. Also, the loss of trees allows for flooding, soil erosion, desertification, and higher temperatures to occur more rapidly and exponentially.

### *2. Increase in Global Warming*

Trees play a major role in controlling global warming. The trees utilize greenhouse gases, restoring the balance in the atmosphere. With constant deforestation, the ratio of greenhouse gases in the atmosphere has increased, adding to our global warming woes.

### *3. Increase in Greenhouse Gas Emissions*

Forests help to mitigate carbon dioxide and other toxic greenhouse gas emissions. However, once they're cut, burned, or otherwise removed, they become carbon sources.

It's estimated that deforestation is responsible for around 20 percent of greenhouse gas emissions, and due to tropical deforestation, 1.5 billion tons of carbon is released every year in the atmosphere.

### *4. Soil Erosion*

Trees are also crucial for our local water cycles as they keep on returning water vapor to the atmosphere. The soil remains moist as the rainwater percolates within the soil.

The fertile soil is held in place by intricate root structures of many layers of trees. With the clearance of tree cover, the land is directly exposed to the sun, making it dry.

Without trees, erosion often occurs and sweeps the land into nearby rivers and streams. Forests serve as nature's water purification plants. Soil erosion makes soil exposed to contaminants that leach into the water supply, which damages the quality of our drinking water.

### *5. Floods*

When it rains, trees absorb and store a large amount of water with the help of their roots. When they are cut down, the flow of water is disrupted, and the soil loses its ability to retain water. It leads to floods in some areas and droughts in others.

### *6. Wildlife Extinction & Habitat Loss*

Due to the massive felling down of trees, various animal species are lost. They lose their habitat and also forced to move to a new location. Many of them are even pushed to extinction.

Our world has lost innumerable species of plants and animals in the last couple of decades. A study of the Brazilian Amazon forecasts that up to 90% of predicted extinctions will occur over the next 40 years.

### *7. Acidic Oceans*

The increased levels of carbon dioxide in the atmosphere due to deforestation and burning fossil fuels make our oceans more acidic. Since the Industrial Revolution, beaches are already 30 percent more acidic, putting ocean species and ecosystems at extreme risk.

### *8. The Decline in Life Quality of People*

People in the millions all over the world depend on forests for hunting, small-scale agriculture, gathering, and medicine. Everyday materials we use, such as latex, cork, fruit, nuts, natural oils, and resins are found in the tropical forests.

Deforestation disrupts the lives of millions of people. In Southeast Asia, deforestation has contributed to social conflict and migration. Poor people from Brazil have been lured from their villages to soy plantations where they are abused and forced at gunpoint to work under inhumane conditions.

### *9. Food Insecurity in the Future*

Deforestation may result in food insecurity in the future. Currently, 52% of all the land used for food production is moderately or severely impacted by soil erosion. In the long term, the lack of fertile soil can lead to low yields and food insecurity.

### *10. Loss of Biodiversity*

Deforestation leads to a huge loss of biodiversity. About 80% of the global biodiversity is located in tropical rainforests. Forests not only provide habitats for wildlife but also foster medicinal conservation.

The forest acts as a critical medium to preserve the wide variety of species. It also destroys the microbial community that is responsible for the production of clean water, the removal of pollutants and the recycling of nutrients.

## Solutions to Deforestation

### *1. Government Regulations*

The best solution to deforestation is to curb the felling of trees by enforcing a series of rules and laws to govern it. Deforestation in the current scenario may have reduced; however, it would be too early to assume.

The money-churner nature of forest resources can be tempting enough for deforestation to continue.

### *2. Banning Clear-Cutting of Forests*

This will curb the total depletion of the forest cover. It is a practical solution and is very feasible.

### *3. Reforestation and Afforestation*

Land skinned of its tree cover for urban settlements should be urged to plant trees in the vicinity and replace the cut trees. Also, the cutting must be replaced by planting young trees to replace the older ones that were cut.

Trees are being planted under several initiatives every year, but they still don't match the numbers of the ones we've already lost.

### *4. Reduce Consumption of Paper*

Your daily consumption of paper includes printing paper, notebooks, napkins, toilet paper, etc. Try to reduce consumption, reduce waste of paper and also opt for recycled paper products.

Make life simple such as printing/writing on both sides of the paper, using less toilet paper, avoiding paper plates and napkins and wherever possible, go paperless.

### *5. Educate Others*

Still, many are entirely unaware of the global warming problem we're facing. Educate your friends, family, and community by sharing the deforestation facts, and its causes and effects. You can make an impact!

### *6. Eat Less Meat*

Livestock rearing has become one of the leading causes of deforestation. Try to eat less meat. It may be hard for some people to try. However, eating less meat, even just for one meal a day, will also make an extreme impact on the environment.

### *7. Purchase from Sustainable, Forest-Friendly Companies*

Try to purchase from companies that are committed to reducing deforestation. Asian Pulp and Paper, Disney, L'Oreal, Hershey, Unilever, Willmar International are deforestation-free.

### *8. Reduce Consumption of Deforestation Prone Products*

Palm oil is a common ingredient in absolutely everything we see around us. Make it a simple habit to get a quick peek at the ingredients. Soybeans are another deforestation hotspot.

Try finding ways to reduce consumption or opting for organic, local soy products and, if possible, avoid it completely.

# Global Agricultural Practices Are a Major Driver of Biodiversity Loss

### *United Nations Environment Programme*

*The United Nations Environment Programme is the global authority that sets the environmental agenda, promotes the coherent implementation of the environmental dimension of sustainable development within the United Nations system, and serves as an authoritative advocate for the global environment.*

"Food System Impacts on Biodiversity Loss," the new Chatham House report, supported by the UN Environment Programme (UNEP) and Compassion in World Farming, describes three actions needed for food system transformation in support of biodiversity, and sets out recommendations to embed food system reform in high level political events over the coming UN "Super Year" for Nature.

Our global food system is the primary driver of biodiversity loss, with agriculture alone being the identified threat to 24,000 of the 28,000 (86%) species at risk of extinction. The global rate of species extinction today is higher than the average rate over the past 10 million years.

In the last decades our food systems have been following the "cheaper food paradigm," with a goal of producing more food at lower costs through increasing inputs such as fertilizers, pesticides, energy, land and water. This paradigm leads to a vicious circle: the lower cost of food production creates a bigger demand for food that must also be produced at a lower cost through more intensification and further land clearance.

The impacts of producing more food at a lower cost are not limited to biodiversity loss. The global food system is a major

"Our Global Food System Is the Primary Driver of Biodiversity Loss," United Nations Environment Programme (UNEP), February 3, 2021. Reprinted by permission.

driver of climate change, accounting for around 30% of total human-produced emissions.

According to the new report, a reform of food systems is a matter of urgency and should focus on three interdependent actions:

Firstly, global dietary patterns need to move towards more plant-heavy diets, mainly due to the disproportionate impact of animal agriculture on biodiversity, land use and the environment. Such a shift, coupled with the reduction of global food waste, would reduce demand and the pressure on the environment and land, benefit the health of populations around the world, and help reduce the risk of pandemics.

Secondly, more land needs to be protected and set aside for nature. The greatest gains for biodiversity will occur when we preserve or restore whole ecosystems. Therefore, we need to avoid converting land for agriculture. Human dietary shifts are essential in order to preserve existing native ecosystems and restore those that have been removed or degraded.

Thirdly, we need to farm in a more nature-friendly, biodiversity-supporting way, limiting the use of inputs and replacing monoculture with polyculture farming practices.

Dietary change is necessary to enable land to be returned to nature, and to allow widespread adoption of nature-friendly farming without increasing the pressure to convert natural land to agriculture. The more the first action is taken up in the form of dietary change, the more scope there is for the second and third actions.

The findings and recommendations of the new Chatham House report were presented today, during an online event which included speakers from UNEP, Chatham House and Compassion in World Farming, as well as Jane Goodall, PhD, DBE, Founder of the Jane Goodall Institute & UN Messenger of Peace. An inspiring panel discussion followed with Louise Mabulo, a chef, environmentalist and UN's Young Champion of the Earth from the Philippines, and Lana Weidgenant, Vice-Chair of Shifting to Sustainable

Consumption Patterns at the UN Food Systems Summit and Deputy Director of Zero Hour International.

## Additional Quotes

Susan Gardner, Director of UNEP's Ecosystems Division, said, "Our current food system is a double edged sword—shaped by decades of the 'cheaper food' paradigm, aimed at producing more food, quickly and cheaply without taking into account the hidden costs to biodiversity and its life-supporting services—and to our own health.

"Reforming the way we produce and consume food is an urgent priority—we need to change global dietary patterns, protect and set aside land for nature and farm in a more nature-friendly and biodiversity-supporting way."

Professor Tim Benton, Research Director, Emerging Risks; Director, Energy, Environment and Resources Programme at Chatham House, said, "The biggest threats to biodiversity arise from exploitative land use—converting natural habitats to agriculture and farming land intensively—and these are driven by the economic demand for producing ever more calorie-rich, but nutritionally poor, food from fewer and fewer commodities grown at scale.

"These commodities underpin a wasteful food system that fails to nourish us and undermines biodiversity and drives climate change."

Philip Lymbery, Global Chief Executive at Compassion in World Farming, said, "At a time when so much of the world continues to battle the Covid-19 pandemic, it's never been more obvious that the well-being of people and animals, wild and farmed, are intertwined.

"As this new report shows, the future of humanity depends on us living in harmony with nature. We need to work with nature, not against her. Never has it been so timely for us to realise that protecting people means protecting animals too.

"The future of farming must be nature-friendly and regenerative, and our diets must become more plant-based, healthy and sustainable.

"Without ending factory farming, we are in danger of having no future at all."

Jane Goodall, PhD, DBE, Founder of the Jane Goodall Institute & UN Messenger of Peace, said, "The intensive farming of billions of animals globally seriously damages the environment, causing loss of biodiversity and producing massive greenhouse gas emissions that accelerate global warming.

"The inhumane crowded conditions not only cause intense suffering to sentient beings but enable the transfer of pathogens from animal to human risking new zoonotic diseases.

"On ethical grounds it should be phased out as soon as possible."

## Humans Are Destroying the Amazon

***Catesby Holmes***

*Catesby Holmes is an editor for The Conversation and works on topics dealing with politics and international issues.*

Nearly 40,000 fires are incinerating Brazil's Amazon rainforest, the latest outbreak in an overactive fire season that has charred 1,330 square miles of the rainforest this year.

Don't blame dry weather for the swift destruction of the world's largest tropical forest, say environmentalists. These Amazonian wildfires are a human-made disaster, set by loggers and cattle ranchers who use a "slash and burn" method to clear land. Feeding off very dry conditions, some of those fires have spread out of control.

Brazil has long struggled to preserve the Amazon, sometimes called the "lungs of the world" because it produces 20% of the world's oxygen. Despite the increasingly strict environmental protections of recent decades, about a quarter of this massive rainforest is already gone—an area the size of Texas.

While climate change endangers the Amazon, bringing hotter weather and longer droughts, development may be the greatest threat facing the rainforest.

Here, environmental researchers explain how farming, big infrastructure projects and roads drive the deforestation that's slowly killing the Amazon.

### 1. Farming in the Jungle

"Deforestation is largely due to land clearing for agricultural purposes, particularly cattle ranching but also soybean production," writes Rachel Garrett, a professor at Boston University who studies land use in Brazil.

"The Amazon Is Burning: 4 Essential Reads on Brazil's Vanishing Rainforest," by Catesby Holmes, The Conversation, August 23, 2019. https://theconversation.com/the-amazon-is-burning-4-essential-reads-on-brazils-vanishing-rainforest-122288. 

Since farmers need "a massive amount of land for grazing," Garrett says, they are driven to "continuously clear forest—illegally—to expand pastureland."

Twelve percent of what was once Amazonian forest—about 93 million acres—is now farmland.

Deforestation in the Amazon has spiked since the election last year of the far-right President Jair Bolsonaro. Arguing that federal conservation zones and hefty fines for cutting down trees hinder economic growth, Bolsonaro has slashed Brazil's strict environmental regulations.

There's no evidence to support Bolsonaro's view, Garrett says.

"Food production in the Amazon has substantially increased since 2004," Garrett says.

The increased production has been pushed by federal policies meant to discourage land clearing, such as hefty fines for deforestation and low-interest loans for investing in sustainable agricultural practices. Farmers are now planting and harvesting two crops—mostly soybean and corn—each year, rather than just one.

Brazilian environmental regulations helped Amazonian ranchers, too.

Garrett's research found that improved pasture management in line with stricter federal land use policies led the number of cattle slaughtered annually per acre to double.

"Farmers are producing more meat—and therefore earning more money—with their land," she writes.

## 2. Infrastructure Development and Deforestation

President Bolsonaro is also pushing forward an ambitious infrastructure development plan that would turn the Amazon's many waterways into electricity generators.

The Brazilian government has long wanted to build a series of big new hydroelectric dams, including on the Tapajós River, the Amazon's only remaining undammed river. But the indigenous Munduruku people, who live near around the Tapajós River, have stridently opposed this idea.

"The Munduruku have until now successfully slowed down and seemingly halted many efforts to profit off the Tapajós," writes Robert T. Walker, a University of Florida professor who has conducted environmental research in the Amazon for 25 years.

But Bolsonaro's government is less likely than his predecessors to respect indigenous rights. One of his first moves in office was to transfer responsibilities for demarcating indigenous lands from the Brazilian Ministry of Justice to the decidedly pro-development Ministry of Agriculture.

And, Walker notes, Bolsonaro's Amazon development plans are part of a broader South American project, conceived in 2000, to build continental infrastructure that provides electricity for industrialization and facilitates trade across the region.

For the Brazilian Amazon, that means not just new dams but also "webs of waterways, rail lines, ports and roads" that will get products like soybeans, corn and beef to market, according to Walker.

"This plan is far more ambitious than earlier infrastructure projects" that damaged the Amazon, Walker writes. If Bolsonaro's plan moves forward, he estimates that fully 40% of the Amazon could be deforested.

## 3. Road-Choked Streams

Roads, most of them dirt, already criss-cross the Amazon.

That came as a surprise to Cecilia Gontijo Leal, a Brazilian researcher who studies tropical fish habitats.

"I imagined that my field work would be all boat rides on immense rivers and long jungle hikes," she writes. "In fact, all my research team needed was a car."

Traveling on rutted mud roads to take water samples from streams across Brazil's Pará state, Leal realized that the informal "bridges" of this locally built transportation network must be impacting Amazonian waterways. So she decided to study that, too.

"We found that makeshift road crossings cause both shore erosion and silt buildup in streams. This worsens water quality, hurting the fish that thrive in this delicately balanced habitat," she writes.

The ill-designed road crossings—which feature perched culverts that disrupt water flow—also act as barriers to movement, preventing fish from finding places to feed, breed and take shelter.

## 4. Rewilding Tropical Forests

The fires now consuming vast swaths of the Amazon are the latest repercussion of development in the Amazon.

Set by farmers likely emboldened by their president's anti-conservation stance, the blazes emit so much smoke that on Aug. 20 it blotted out the midday sun in the city of São Paulo, 1,700 miles away. The fires are still multiplying, and peak dry season is still a month away.

Apocalyptic as this sounds, science suggests it's not too late to save the Amazon.

Tropical forests destroyed by fire, logging, land-clearing and roads can be replanted, say ecologists Robin Chazdon and Pedro Brancalion.

Using satellite imagery and the latest peer-reviewed research on biodiversity, climate change and water security, Chazdon and Brancalion identified 385,000 square miles of "restoration hotspots"—areas where restoring tropical forests would be most beneficial, least costly and lowest risk.

"Although these second-growth forests will never perfectly replace the older forests that have been lost," Chazon writes, "planting carefully selected trees and assisting natural recovery processes can restore many of their former properties and functions."

The five countries with the most tropical restoration potential are Brazil, Indonesia, India, Madagascar and Colombia.

Current Controversies

Chapter 3

# Are Changes in Climate a Natural Phenomenon?

# Overview: Natural Climate Altering Phenomenon Explained

*Howard Lee*

*Howard Lee is a geologist and science writer focusing on Earth's climate through deep time.*

Earth has been a snowball and a hothouse at different times in its past. So if the climate changed before humans, how can we be sure we're responsible for the dramatic warming that's happening today?

In part it's because we can clearly show the causal link between carbon dioxide emissions from human activity and the 1.28 degree Celsius (and rising) global temperature increase since preindustrial times. Carbon dioxide molecules absorb infrared radiation, so with more of them in the atmosphere, they trap more of the heat radiating off the planet's surface below.

But paleoclimatologists have also made great strides in understanding the processes that drove climate change in Earth's past. Here's a primer on 10 ways climate varies naturally, and how each compares with what's happening now.

## Solar Cycles

**Magnitude:** 0.1 to 0.3 degrees Celsius of cooling
**Time frame:** 30- to 160-year downturns in solar activity separated by centuries

Every 11 years, the sun's magnetic field flips, driving an 11-year cycle of solar brightening and dimming. But the variation is small and has a negligible impact on Earth's climate.

More significant are "grand solar minima," decades-long periods of reduced solar activity that have occurred 25 times in the last 11,000 years. A recent example, the Maunder minimum, which

"How Earth's Climate Changes Naturally (and Why Things Are Different Now)," by Howard Lee, *Quanta Magazine*, July 21, 2020. Reprinted by permission.

occurred between 1645 and 1715, saw solar energy drop by 0.04% to 0.08% below the modern average. Scientists long thought the Maunder minimum might have caused the "Little Ice Age," a cool period from the 15th to the 19th century; they've since shown it was too small and occurred at the wrong time to explain the cooling, which probably had more to do with volcanic activity.

The sun has been dimming slightly for the last half-century while the Earth heats up, so global warming cannot be blamed on the sun.

## Volcanic Sulfur

**Magnitude:** Approximately 0.6 to 2 degrees Celsius of cooling
**Time frame:** 1 to 20 years

In the year 539 or 540 A.D., the Ilopango volcano in El Salvador exploded so violently that its eruption plume reached high into the stratosphere. Cold summers, drought, famine and plague devastated societies around the world.

Eruptions like Ilopango's inject the stratosphere with reflective droplets of sulfuric acid that screen sunlight, cooling the climate. Sea ice can increase as a result, reflecting more sunlight back to space and thereby amplifying and prolonging the global cooling.

Ilopango triggered a roughly 2 degree Celsius drop that lasted 20 years. More recently, the eruption of Pinatubo in the Philippines in 1991 cooled the global climate by 0.6 degrees Celsius for 15 months.

Volcanic sulfur in the stratosphere can be disruptive, but in the grand scale of Earth's history it's tiny and temporary.

## Short-Term Climate Fluctuations

**Magnitude:** Up to 0.15 degrees Celsius
**Time frame:** 2 to 7 years

On top of seasonal weather patterns, there are other short-term cycles that affect rainfall and temperature. The most significant, the El Niño–Southern Oscillation, involves circulation changes in the tropical Pacific Ocean on a time frame of two to seven years that strongly influence rainfall in North America. The North Atlantic

Oscillation and the Indian Ocean Dipole also produce strong regional effects. Both of these interact with the El Niño–Southern Oscillation.

The interconnections between these cycles used to make it hard to show that human-caused climate change was statistically significant and not just another lurch of natural variability. But anthropogenic climate change has since gone well beyond natural variability in weather and seasonal temperatures. The U.S. National Climate Assessment in 2017 concluded that there's "no convincing evidence for natural cycles in the observational record that could explain the observed changes in climate."

## Orbital Wobbles

**Magnitude:** Approximately 6 degrees Celsius in the last 100,000-year cycle; varies through geological time
**Time frame:** Regular, overlapping cycles of 23,000, 41,000, 100,000, 405,000 and 2,400,000 years

Earth's orbit wobbles as the sun, the moon and other planets change their relative positions. These cyclical wobbles, called Milankovitch cycles, cause the amount of sunlight to vary at middle latitudes by up to 25% and cause the climate to oscillate. These cycles have operated throughout time, yielding the alternating layers of sediment you see in cliffs and road cuts.

During the Pleistocene epoch, which ended about 11,700 years ago, Milankovitch cycles sent the planet in and out of ice ages. When Earth's orbit made northern summers warmer than average, vast ice sheets across North America, Europe and Asia melted; when the orbit cooled northern summers, those ice sheets grew again. Since warmer oceans dissolve less carbon dioxide, atmospheric carbon dioxide levels rose and fell in concert with these orbital wobbles, amplifying their effects.

Today Earth is approaching another minimum of northern sunlight, so without human carbon dioxide emissions we would be heading into another ice age within the next 1,500 years or so.

## Faint Young Sun

**Magnitude:** No net temperature effect
**Time frame:** Constant

Though the sun's brightness fluctuates on shorter timescales, it brightens overall by 0.009% per million years, and it has brightened by 48% since the birth of the solar system 4.5 billion years ago.

Scientists reason that the faintness of the young sun should have meant that Earth remained frozen solid for the first half of its existence. But, paradoxically, geologists have found 3.4-billion-year-old rocks that formed in wave-agitated water. Earth's unexpectedly warm early climate is probably explained by some combination of less land erosion, clearer skies, a shorter day and a peculiar atmospheric composition before Earth had an oxygen-rich atmosphere.

Clement conditions in the second half of Earth's existence, despite a brightening sun, do not create a paradox: Earth's weathering thermostat counteracts the effects of the extra sunlight, stabilizing Earth's temperature (see next section).

## Carbon Dioxide and the Weathering Thermostat

**Magnitude:** Counteracts other changes
**Time frame:** 100,000 years or longer

The main control knob for Earth's climate through deep time has been the level of carbon dioxide in the atmosphere, since carbon dioxide is a long-lasting greenhouse gas that blocks heat that tries to rise off the planet.

Volcanoes, metamorphic rocks and the oxidization of carbon in eroded sediments all emit carbon dioxide into the sky, while chemical reactions with silicate minerals remove carbon dioxide and bury it as limestone. The balance between these processes works as a thermostat, because when the climate warms, chemical reactions become more efficient at removing carbon dioxide, putting a brake on the warming. When the climate cools, reactions become less efficient, easing the cooling. Consequently, over the very long term, Earth's climate has remained relatively stable, providing a habitable environment. In particular, average carbon dioxide levels have declined steadily in response to solar brightening.

However, the weathering thermostat takes hundreds of thousands of years to react to changes in atmospheric carbon dioxide. Earth's oceans can act somewhat faster to absorb and remove excess carbon, but even that takes millennia and can be overwhelmed, leading to ocean acidification. Each year, the burning of fossil fuels emits about 100 times more carbon dioxide than volcanoes emit—too much too fast for oceans and weathering to neutralize it, which is why our climate is warming and our oceans are acidifying.

## Plate Tectonics

**Magnitude:** Roughly 30 degrees Celsius over the past 500 million years
**Time frame:** Millions of years

The rearrangement of land masses on Earth's crust can slowly shift the weathering thermostat to a new setting.

The planet has generally been cooling for the last 50 million years or so, as plate tectonic collisions thrust up chemically reactive rock like basalt and volcanic ash in the warm, wet tropics, increasing the rate of reactions that draw carbon dioxide from the sky. Additionally, over the last 20 million years, the building of the Himalayas, Andes, Alps and other mountains has more than doubled erosion rates, boosting weathering. Another contributor to the cooling trend was the drifting apart of South America and Tasmania from Antarctica 35.7 million years ago, which initiated a new ocean current around Antarctica. This invigorated ocean circulation and carbon dioxide–consuming plankton; Antarctica's ice sheets subsequently grew substantially.

Earlier, in the Jurassic and Cretaceous periods, dinosaurs roamed Antarctica because enhanced volcanic activity, in the absence of those mountain chains, sustained carbon dioxide levels around 1,000 parts per million, compared to 415 ppm today. The average temperature of this ice-free world was 5 to 9 degrees Celsius warmer than now, and sea levels were around 250 feet higher.

## Asteroid Impacts

**Magnitude:** Approximately 20 degrees Celsius of cooling followed by 5 degrees Celsius of warming (Chicxulub)
**Time frame:** Centuries of cooling, 100,000 years of warming (Chicxulub)

The Earth Impact Database recognizes 190 craters with confirmed impact on Earth so far. None had any discernable effect on Earth's climate except for the Chicxulub impact, which vaporized part of Mexico 66 million years ago, killing off the dinosaurs. Computer modeling suggests that Chicxulub blasted enough dust and sulfur into the upper atmosphere to dim sunlight and cool Earth by more than 20 degrees Celsius, while also acidifying the oceans. The planet took centuries to return to its pre-impact temperature, only to warm by a further 5 degrees Celsius, due to carbon dioxide in the atmosphere from vaporized Mexican limestone.

How or whether volcanic activity in India around the same time as the impact exacerbated the climate change and mass extinction remains controversial.

## Evolutionary Changes

**Magnitude:** Depends on event; about 5 degrees Celsius cooling in late Ordovician (445 million years ago)
**Time frame:** Millions of years

Occasionally, the evolution of new kinds of life has reset Earth's thermostat. Photosynthetic cyanobacteria that arose some 3 billion years ago, for instance, began terraforming the planet by emitting oxygen. As they proliferated, oxygen eventually rose in the atmosphere 2.4 billion years ago, while methane and carbon dioxide levels plummeted. This plunged Earth into a series of "snowball" climates for 200 million years. The evolution of ocean life larger than microbes initiated another series of snowball climates 717 million years ago—in this case, it was because the organisms began raining detritus into the deep ocean, exporting carbon from the atmosphere into the abyss and ultimately burying it.

When the earliest land plants evolved about 230 million years later in the Ordovician period, they began forming the terrestrial biosphere, burying carbon on continents and extracting land nutrients that washed into the oceans, boosting life there, too. These changes probably triggered the ice age that began about 445 million years ago. Later, in the Devonian period, the evolution of trees further reduced carbon dioxide and temperatures, conspiring with mountain building to usher in the Paleozoic ice age.

## Large Igneous Provinces

**Magnitude:** Around 3 to 9 degrees Celsius of warming
**Time frame:** Hundreds of thousands of years

Continent-scale floods of lava and underground magma called large igneous provinces have ushered in many of Earth's mass extinctions. These igneous events unleashed an arsenal of killers (including acid rain, acid fog, mercury poisoning and destruction of the ozone layer), while also warming the planet by dumping huge quantities of methane and carbon dioxide into the atmosphere more quickly than the weathering thermostat could handle.

In the end-Permian event 252 million years ago, which wiped out 81% of marine species, underground magma ignited Siberian coal, drove up atmospheric carbon dioxide to 8,000 parts per million and raised the temperature by between 5 and 9 degrees Celsius. The more minor Paleocene-Eocene Thermal Maximum event 56 million years ago cooked methane in North Atlantic oil deposits and funneled it into the sky, warming the planet by 5 degrees Celsius and acidifying the ocean; alligators and palms subsequently thrived on Arctic shores. Similar releases of fossil carbon deposits happened in the end-Triassic and the early Jurassic; global warming, ocean dead zones and ocean acidification resulted.

As a team of researchers studying the end-Triassic event wrote in April in Nature Communications, "Our estimates suggest that the amount of CO2 that each ... magmatic pulse injected into the end-Triassic atmosphere is comparable to the amount of anthropogenic emissions projected for the 21st century."

# The "Little Ice Age" Caused Death and Destruction

### *Dagomar Degroot*

*As a professor of environmental history at Georgetown University, Dagomar Degroot writes and teaches about the history of climate change.*

In recent weeks, catastrophic floods overwhelmed towns in Germany, Belgium and the Netherlands, inundated subway tunnels in China, swept through northwestern Africa and triggered deadly landslides in India and Japan. Heat and drought fanned wildfires in the North American West and Siberia, contributed to water shortages in Iran, and worsened famines in Ethiopia, Somalia and Kenya.

Extremes like these are increasingly caused or worsened by human activities heating up Earth's climate. For thousands of years, Earth's climate has not changed anywhere near as quickly or profoundly as it's changing today.

Yet on a smaller scale, humans have seen waves of extreme weather events coincide with temperature changes before. It happened during what's known as the Little Ice Age, a period between the 14th and 19th centuries that was marked by large volcanic eruptions and bitter cold spells in parts of the world.

The global average temperature is believed to have cooled by less than a half-degree Celsius (less than 0.9 F) during even the chilliest decades of the Little Ice Age, but locally, extremes were common.

In diaries and letters from that period, people wrote about "years without a summer," when wintry weather persisted long after spring. In one such summer, in 1816, cold that followed a

"Small Climate Changes Can Have Devastating Local Consequences—It Happened in the Little Ice Age," by Dagomar Degroot, The Conversation, July 27, 2021. https://theconversation.com/small-climate-changes-can-have-devastating-local-consequences-it-happened-in-the-little-ice-age-164916. 

massive volcanic eruption in Indonesia ruined crops across parts of Europe and North America. Less well known are the unusually cold European summers of 1587, 1628 and 1675, when unseasonal frost provoked fear and, in some places, hunger.

"It is horribly cold," author Marie de Rabutin-Chantal wrote from Paris during the last of these years; "the behavior of the sun and of the seasons has changed."

Winters could be equally terrifying. People reported 17th-century blizzards as far south as Florida and the Chinese province of Fujian. Sea ice trapped ships, repeatedly enclosed the Chesapeake Bay and froze over rivers from the Bosporus to the Meuse. In early 1658, ice so completely covered the Baltic Sea that a Swedish army marched across the water separating Sweden and Denmark to besiege Copenhagen. Poems and songs suggest people simply froze to death while huddling in their homes.

These were cold snaps, not heat waves, but the overall story should seem familiar: A small global change in climate dramatically altered the likelihood of extreme local weather. Scholars who study the history of climate and society, like me, identify these changes in the past and find out how human populations responded.

## What's Behind the Extremes

We know about the Little Ice Age because the natural world is full of things like trees, stalagmites and ice sheets that respond to weather while growing or accumulating gradually over time. Specialists can use past fluctuations in their growth or chemistry as indicators of fluctuations in climate and thereby create graphs or maps—reconstructions—that show historical climate changes.

These reconstructions reveal that waves of cooling swept across much of the world. They also suggest likely causes—including a series of explosive volcanic eruptions that abruptly released sunlight-scattering dust into the stratosphere; and slow, internal variability in regional patterns of atmospheric and oceanic circulation.

These causes could only cool the Earth by a few tenths of a degree Celsius during the chilliest waves of the Little Ice

Age, however. And the cooling was not nearly as consistent as present-day warming.

Small global trends can mask far bigger local changes. Studies have suggested that modest cooling created by volcanic eruptions can reduce the usual contrast between temperatures over land and sea, because land heats and cools faster than oceans. Since that contrast powers the monsoons, the African and East Asian summer monsoons can weaken after big eruptions. That likely disturbed atmospheric circulation all the way into the North Atlantic, reducing the flow of warm air into Europe. This is why parts of Western Europe, for example, may have cooled by more than 3 C (5.4 F) even as the rest of the world cooled far less during the 1816 year without a summer.

Feedback loops also amplified and sustained regional cooling, similar to how they amplify regional warming today. In the Arctic, for example, cooler temperatures can mean more, longer-lasting sea ice. Ice reflects more sunlight back into space than water does, and that feedback loop leads to more cooling, more ice and so on. As a result, the comparatively modest climate changes of the Little Ice Age likely had profound local impacts.

Changing patterns of atmospheric circulation and pressure also led in many regions to remarkably wet, dry or stormy weather.

Heavy sea ice in the Greenland Sea may have diverted the North Atlantic storm track south, funneling severe gales toward the dikes and dams of what are today the Netherlands and Belgium. Thousands of people succumbed in the 1570 All Saints' Day Flood along the German and Dutch coast, and again in the Christmas Flood of 1717. Heavy precipitation and water pooling behind dams of melting ice repeatedly overwhelmed inadequate flood defenses and inundated central and Western Europe. "Who would not take pity on the city?" one chronicler lamented after seeing his town under water and then on fire in 1602. "One storm, one flood, one fire destroyed it all."

Cooling sea surface temperatures in the North Atlantic Ocean probably also diverted the rain-giving winds around the equator to the south, provoking droughts that undermined the water infrastructure of 15th-century Angkor.

Owing perhaps to the modest cooling of volcanic dust veils, disrupted patterns of atmospheric circulation led in the 16th century to severe droughts that contributed to food shortages across the Ottoman Empire. In 1640, the grand canal that supplied Beijing with food simply dried up, and a short but profound drought in 1666 primed the wooden infrastructure of European cities for a wave of catastrophic urban fires.

## How Does It Apply to Today?

Today, the temperature shift is going in the other direction—with global temperatures already 1 C (1.8 F) higher than before the industrial era, and local, sometimes devastating, extremes occurring around the world.

New research has found that extreme heat waves, those that don't just break records but shatter them, become more common when temperatures change rapidly.

These serve as a warning to governments to redouble their efforts to limit warming to 1.5 C (2.7 F), relative to the 20th-century average, while also investing in the development and deployment of technologies that filter greenhouse gases out of the atmosphere.

Restoring the chemistry of the atmosphere will still take many decades after countries bring down their greenhouse gas emissions, and so communities must adapt to a hotter and less habitable planet. Nations and communities might learn from some of the success stories of the Little Ice Age: Populations that prospered were often those that provided for their poor, established diverse trade networks, migrated from vulnerable environments, and above all adapted proactively to new environmental realities.

People who lived through the Little Ice Age lacked perhaps the most important resource available today: the ability to learn from the long global history of human responses to climate change.

# Earth's Orbital Cycles Influence Climate

*Alan Buis*

*Alan Buis is a science writer at the NASA Jet Propulsion Laboratory.*

Our lives literally revolve around cycles: series of events that are repeated regularly in the same order. There are hundreds of different types of cycles in our world and in the universe. Some are natural, such as the change of the seasons, annual animal migrations or the circadian rhythms that govern our sleep patterns. Others are human-produced, like growing and harvesting crops, musical rhythms or economic cycles.

Cycles also play key roles in Earth's short-term weather and long-term climate. A century ago, Serbian scientist Milutin Milankovitch hypothesized the long-term, collective effects of changes in Earth's position relative to the Sun are a strong driver of Earth's long-term climate, and are responsible for triggering the beginning and end of glaciation periods (Ice Ages).

Specifically, he examined how variations in three types of Earth orbital movements affect how much solar radiation (known as insolation) reaches the top of Earth's atmosphere as well as where the insolation reaches. These cyclical orbital movements, which became known as the Milankovitch cycles, cause variations of up to 25 percent in the amount of incoming insolation at Earth's mid-latitudes (the areas of our planet located between about 30 and 60 degrees north and south of the equator).

The Milankovitch cycles include:

1. The shape of Earth's orbit, known as eccentricity;
2. The angle Earth's axis is tilted with respect to Earth's orbital plane, known as obliquity; and
3. The direction Earth's axis of rotation is pointed, known as precession.

"Milankovitch (Orbital) Cycles and Their Role in Earth's Climate," by Alan Buis, NASA's Global Climate Change, February 27, 2020.

Let's take a look at each.

## Eccentricity

Earth's annual pilgrimage around the Sun isn't perfectly circular, but it's pretty close. Over time, the pull of gravity from our solar system's two largest gas giant planets, Jupiter and Saturn, causes the shape of Earth's orbit to vary from nearly circular to slightly elliptical. Eccentricity measures how much the shape of Earth's orbit departs from a perfect circle. These variations affect the distance between Earth and the Sun.

Eccentricity is the reason why our seasons are slightly different lengths, with summers in the Northern Hemisphere currently about 4.5 days longer than winters, and springs about three days longer than autumns. As eccentricity decreases, the length of our seasons gradually evens out.

The difference in the distance between Earth's closest approach to the Sun (known as perihelion), which occurs on or about January 3 each year, and its farthest departure from the Sun (known as aphelion) on or about July 4, is currently about 5.1 million kilometers (about 3.2 million miles), a variation of 3.4 percent. That means each January, about 6.8 percent more incoming solar radiation reaches Earth than it does each July.

When Earth's orbit is at its most elliptic, about 23 percent more incoming solar radiation reaches Earth at our planet's closest approach to the Sun each year than does at its farthest departure from the Sun. Currently, Earth's eccentricity is near its least elliptic (most circular) and is very slowly decreasing, in a cycle that spans about 100,000 years.

The total change in global annual insolation due to the eccentricity cycle is very small. Because variations in Earth's eccentricity are fairly small, they're a relatively minor factor in annual seasonal climate variations.

## Obliquity

The angle Earth's axis of rotation is tilted as it travels around the Sun is known as obliquity. Obliquity is why Earth has seasons. Over the last million years, it has varied between 22.1 and 24.5 degrees perpendicular to Earth's orbital plane. The greater Earth's axial tilt angle, the more extreme our seasons are, as each hemisphere receives more solar radiation during its summer, when the hemisphere is tilted toward the Sun, and less during winter, when it is tilted away. Larger tilt angles favor periods of deglaciation (the melting and retreat of glaciers and ice sheets). These effects aren't uniform globally—higher latitudes receive a larger change in total solar radiation than areas closer to the equator.

Earth's axis is currently tilted 23.4 degrees, or about half way between its extremes, and this angle is very slowly decreasing in a cycle that spans about 41,000 years. It was last at its maximum tilt about 10,700 years ago and will reach its minimum tilt about 9,800 years from now. As obliquity decreases, it gradually helps make our seasons milder, resulting in increasingly warmer winters, and cooler summers that gradually, over time, allow snow and ice at high latitudes to build up into large ice sheets. As ice cover increases, it reflects more of the Sun's energy back into space, promoting even further cooling.

## Precession

As Earth rotates, it wobbles slightly upon its axis, like a slightly off-center spinning toy top. This wobble is due to tidal forces caused by the gravitational influences of the Sun and Moon that cause Earth to bulge at the equator, affecting its rotation. The trend in the direction of this wobble relative to the fixed positions of stars is known as axial precession. The cycle of axial precession spans about 25,771.5 years.

Axial precession makes seasonal contrasts more extreme in one hemisphere and less extreme in the other. Currently perihelion occurs during winter in the Northern Hemisphere and in summer in the Southern Hemisphere. This makes Southern Hemisphere summers hotter and moderates Northern Hemisphere seasonal variations. But in about 13,000 years, axial precession will cause these conditions to flip, with the Northern Hemisphere seeing more extremes in solar radiation and the Southern Hemisphere experiencing more moderate seasonal variations.

Axial precession also gradually changes the timing of the seasons, causing them to begin earlier over time, and gradually changes which star Earth's axis points to at the North Pole (the North Star). Today Earth's North Stars are Polaris and Polaris Australis, but a couple of thousand years ago, they were Kochab and Pherkad.

There's also apsidal precession. Not only does Earth's axis wobble, but Earth's entire orbital ellipse also wobbles irregularly, primarily due to its interactions with Jupiter and Saturn. The cycle of apsidal precession spans about 112,000 years. Apsidal precession changes the orientation of Earth's orbit relative to the elliptical plane.

The combined effects of axial and apsidal precession result in an overall precession cycle spanning about 23,000 years on average.

## A Climate Time Machine

The small changes set in motion by Milankovitch cycles operate separately and together to influence Earth's climate over very long timespans, leading to larger changes in our climate over tens of thousands to hundreds of thousands of years. Milankovitch combined the cycles to create a comprehensive mathematical model for calculating differences in solar radiation at various Earth latitudes along with corresponding surface temperatures. The model is sort of like a climate time machine: it can be run backward and forward to examine past and future climate conditions.

Milankovitch assumed changes in radiation at some latitudes and in some seasons are more important than others to the growth and retreat of ice sheets. In addition, it was his belief that obliquity was the most important of the three cycles for climate, because it affects the amount of insolation in Earth's northern high-latitude regions during summer (the relative role of precession versus obliquity is still a matter of scientific study).

He calculated that Ice Ages occur approximately every 41,000 years. Subsequent research confirms that they did occur at 41,000-year intervals between one and three million years ago. But about 800,000 years ago, the cycle of Ice Ages lengthened to 100,000 years, matching Earth's eccentricity cycle. While various theories have been proposed to explain this transition, scientists do not yet have a clear answer.

Milankovitch's work was supported by other researchers of his time, and he authored numerous publications on his hypothesis. But it wasn't until about 10 years after his death in 1958 that the global science community began to take serious notice of his theory. In 1976, a study in the journal *Science* by Hays et al. using deep-sea sediment cores found that Milankovitch cycles correspond with periods of major climate change over the past 450,000 years, with Ice Ages occurring when Earth was undergoing different stages of orbital variation.

Several other projects and studies have also upheld the validity of Milankovitch's work, including research using data from ice cores in Greenland and Antarctica that has provided strong evidence of Milankovitch cycles going back many hundreds of thousands of years. In addition, his work has been embraced by the National Research Council of the U.S. National Academy of Sciences.

Scientific research to better understand the mechanisms that cause changes in Earth's rotation and how specifically Milankovitch cycles combine to affect climate is ongoing. But the theory that they drive the timing of glacial-interglacial cycles is well accepted.

# Man-Made Activity Is Increasing the Natural Greenhouse Effect

*Melissa Denchak*

*Melissa Denchak is a freelance editor and writer for the Natural Resources Defense Council.*

The greenhouse effect is a good thing. It warms the planet to its comfortable average of 59 degrees Fahrenheit (15 degrees Celsius) and keeps life on earth, well, livable. Without it the world would be a frozen, uninhabitable place, more like Mars. The problem is, mankind's voracious burning of fossil fuels for energy is artificially amping up the natural greenhouse effect. The result? An increase in global warming that is altering the planet's climate systems in countless ways. Here's a look at what the greenhouse effect is, what causes it, and how we can temper its contributions to our changing climate.

## What Is the Greenhouse Effect?

Identified by scientists as far back as 1896, the greenhouse effect is the natural warming of the earth that results when gases in the atmosphere trap heat from the sun that would otherwise escape into space.

## What Causes the Greenhouse Effect?

Sunlight makes the earth habitable. While 30 percent of the solar energy that reaches our world is reflected back to space, approximately 70 percent passes through the atmosphere to the earth's surface, where it is absorbed by the land, oceans, and atmosphere, and heats the planet. This heat is then radiated back up in the form of invisible infrared light. While some of this infrared light continues on into space, the vast majority—indeed,

"Greenhouse Effect 101," by Melissa Denchak, NRDC (Natural Resources Defense Council), July 16, 2019. Reprinted by permission.

some 90 percent—gets absorbed by atmospheric gases, known as greenhouse gases, and redirected back toward the earth, causing further warming.

For most of the past 800,000 years—much longer than human civilization has existed—the concentration of greenhouse gases in our atmosphere was between about 200 and 280 parts per million. (In other words, there were 200 to 280 molecules of the gases per million molecules of air.) But in the past century, that concentration has jumped to more than 400 parts per million, driven up by human activities such as burning fossil fuels and deforestation. The higher concentrations of greenhouse gases—and carbon dioxide in particular—is causing extra heat to be trapped and global temperatures to rise.

## What Are Greenhouse Gases?

Earth's greenhouse gases trap heat in the atmosphere and warm the planet. The main gases responsible for the greenhouse effect include carbon dioxide, methane, nitrous oxide, and water vapor (which all occur naturally), and fluorinated gases (which are synthetic). Greenhouse gases have different chemical properties and are removed from the atmosphere, over time, by different processes. Carbon dioxide, for example, is absorbed by so-called carbon sinks such as plants, soil, and the ocean. Fluorinated gases are destroyed only by sunlight in the far upper atmosphere.

How much any one greenhouse gas influences global warming depends on three key factors. The first is how much of it exists in the atmosphere. Concentrations are measured in parts per million (ppm), parts per billion (ppb), or parts per trillion (ppt); 1 ppm for a given gas means, for example, that there is one molecule of that gas in every 1 million molecules of air. The second is its lifetime—how long it remains in the atmosphere. The third is how effective it is at trapping heat. This is referred to as its global warming potential, or GWP, and is a measure of the total energy that a gas absorbs over a given period of time (usually 100 years) relative to the emissions of 1 ton of carbon dioxide.

Radiative forcing (RF) is another way to measure greenhouse gases (and other climate drivers, such as the sun's brightness and large volcanic eruptions). Also known as climate forcing, RF quantifies the difference between how much of the sun's energy gets absorbed by the earth and how much is released into space as a result of any one climate driver. A climate driver with a positive RF value indicates that it has a warming effect on the planet; a negative value represents cooling.

## What Are Greenhouse Gas Emissions?

Since the start of the Industrial Revolution and the advent of coal-powered steam engines, human activities have vastly increased the volume of greenhouse gases emitted into the atmosphere. It is estimated that between 1750 and 2011, atmospheric concentrations of carbon dioxide increased by 40 percent, methane by 150 percent, and nitrous oxide by 20 percent. In the late 1920s, we started adding man-made fluorinated gases like chlorofluorocarbons, or CFCs, to the mix.

In recent decades we've only picked up the pace. Of all the man-made emissions of carbon dioxide—the most abundant greenhouse gas released by human activities, and one of the longest-lasting—from 1750 to 2010, approximately half were generated in the last 40 years alone, in large part due to fossil fuel combustion and industrial processes. And while global greenhouse gas emissions have occasionally plateaued or dropped from year to year (most recently between 2014 and 2016), they're accelerating once again. In 2017, carbon emissions rose by 1.6 percent; in 2018 they increased by an estimated 2.7 percent.

## Five Major Greenhouse Gases

The most significant gases that cause global warming via the greenhouse effect are the following:

### *Carbon Dioxide*

Accounting for about 76 percent of global human-caused emissions, carbon dioxide ($CO_2$) sticks around for quite a while. Once it's emitted into the atmosphere, 40 percent still remains after 100 years, 20 percent after 1,000 years, and 10 percent as long as 10,000 years later.

### *Methane*

Although methane ($CH_4$) persists in the atmosphere for far less time than carbon dioxide (about a decade), it is much more potent in terms of the greenhouse effect. In fact, pound for pound, its global warming impact is 25 times greater than that of carbon dioxide over a 100-year period. Globally it accounts for approximately 16 percent of human-generated greenhouse gas emissions.

### *Nitrous Oxide*

Nitrous oxide ($N_2O$) is a powerful greenhouse gas: It has a GWP 300 times that of carbon dioxide on a 100-year time scale, and it remains in the atmosphere, on average, a little more than a century. It accounts for about 6 percent of human-caused greenhouse gas emissions worldwide.

### *Fluorinated Gases*

Emitted from a variety of manufacturing and industrial processes, fluorinated gases are man-made. There are four main categories: hydrofluorocarbons (HFCs), perfluorocarbons (PFCs), sulfur hexafluoride ($SF_6$), and nitrogen trifluoride ($NF_3$).

Although fluorinated gases are emitted in smaller quantities than other greenhouse gases (they account for just 2 percent of man-made global greenhouse gas emissions), they trap substantially more heat. Indeed, the GWP for these gases can be in the thousands to tens of thousands, and they have long atmospheric lifetimes, in some cases lasting tens of thousands of years.

HFCs are used as a replacement for ozone-depleting chlorofluorocarbons (CFCs) and hydrochlorofluorocarbons (HCFCs), usually in air conditioners and refrigerators, but some are being phased out because of their high GWP. Replacing these HFCs and properly disposing of them is considered to be one of the most important climate actions the world can take.

### *Water Vapor*

The most abundant greenhouse gas overall, water vapor differs from other greenhouse gases in that changes in its atmospheric concentrations are linked not to human activities directly, but rather to the warming that results from the other greenhouse gases we emit. Warmer air holds more water. And since water vapor is a greenhouse gas, more water absorbs more heat, inducing even greater warming and perpetuating a positive feedback loop. (It's worth noting, however, that the net impact of this feedback loop is still uncertain, as increased water vapor also increases cloud cover that reflects the sun's energy away from the earth.)

## Where Do Greenhouse Gases Come From?

Population size, economic activity, lifestyle, energy use, land use patterns, technology, and climate policy: According to the Intergovernmental Panel on Climate Change (IPCC), these are the broad forcings that drive nearly all human-caused greenhouse gas emissions. Here's a closer look at greenhouse gas emissions by source.

### *Electricity and Heat Production*

The burning of coal, oil, and natural gas to produce electricity and heat accounts for one-quarter of worldwide human-driven emissions, making it the largest single source. In the United States it's the second-largest (behind transportation), responsible for about 27.5 percent of U.S. emissions in 2017, with carbon dioxide the primary gas released (along with small amounts of methane and nitrous oxide), mainly from coal combustion.

### *Agriculture and Land Use Changes*

About another quarter of global greenhouse gas emissions stem from agriculture and other land-use activities (such as deforestation). In the United States, agricultural activities—primarily the raising of livestock and crops for food—accounted for 8.4 percent of greenhouse gas emissions in 2017. Of those, the vast majority were methane (which is produced as manure decomposes and as beef and dairy cows belch and pass gas) and nitrous oxide (often released with the use of nitrogen-heavy fertilizers).

Trees, plants, and soil absorb carbon dioxide from the air. The plants and trees do it via photosynthesis (a process by which they turn carbon dioxide into glucose); the soil houses microbes that carbon binds to. So nonagricultural land-use changes such as deforestation, reforestation (replanting in existing forested areas), and afforestation (creating new forested areas) can either increase the amount of carbon in the atmosphere (as in the case of deforestation) or decrease it via absorption, removing more carbon dioxide from the air than they emit. (When trees or plants are cut down, they no longer absorb carbon dioxide, and when they are burned or decompose, they release carbon dioxide back into the atmosphere.) In the United States, land-use activities currently represent a net carbon sink, absorbing more carbon dioxide from the air than they emit.

### *Industry*

About one-fifth of global human-driven emissions come from the industrial sector, which includes the manufacturing of goods and raw materials (like cement and steel), food processing, and construction. In 2017, industry accounted for 22.4 percent of U.S. man-made emissions, of which the majority was carbon dioxide, though methane, nitrous oxide, and fluorinated gases were also released.

### Transportation

The burning of petroleum-based fuels, namely gasoline and diesel, to power the world's transportation systems accounts for 14 percent of global greenhouse gas emissions. In the United States, with Americans buying larger cars and taking more flights and with low gas prices encouraging drivers to use their cars more, transportation is the largest contributor of greenhouse gases. (It accounted for 28.7 percent of U.S. emissions in 2017.) Carbon dioxide is the primary gas emitted, though fuel combustion also releases small amounts of methane and nitrous oxide, and vehicle air conditioning and refrigerated transport release fluorinated gases too.

Nationwide, cars and trucks are responsible for more than 80 percent of transportation-related carbon emissions.

### Buildings

Operating buildings around the world generates 6.4 percent of global greenhouse gases. In the United States, homes and businesses accounted for about 11 percent of warming emissions. These emissions, made up mostly of carbon dioxide and methane, stem primarily from burning natural gas and oil for heating and cooking, though other sources include managing waste and wastewater and leaking refrigerants from air-conditioning and refrigeration systems.

### Other Sources

This category includes emissions from energy-related activities other than fossil fuel combustion, such as the extraction, refining, processing, and transportation of oil, gas, and coal. Globally, this sector accounts for 9.6 percent of emissions.

## Greenhouse Gas Emissions by Country

Since the start of the Industrial Revolution, more than 2,000 billion tons of carbon dioxide have been released into the atmosphere by human activities, according to the Global Carbon Project. North America and Europe are responsible for approximately half of that total, while the emerging economies of China and India have contributed another 14 percent. For the remainder, 150-plus countries share responsibility.

An analysis of carbon dioxide emissions by country today shows that China now leads the pack, responsible for 27 percent of all emissions. Next comes the United States (15 percent), the European Union's 28 member states including the United Kingdom (10 percent), and India (7 percent) next. Together, these global powers account for almost 60 percent of all emissions.

## The Consequences of the Greenhouse Effect

Today's human-caused greenhouse gas emissions are higher than ever, the concentration of greenhouse gases in the atmosphere is rising rapidly, and according to the IPCC, the planet is heating up. Between preindustrial times and now, the earth's average temperature has increased 1.8 degrees Fahrenheit (1.0 degrees Celsius), with approximately two-thirds of that warming occurring in the last handful of decades alone. According to the IPCC, 1983 to 2012 was likely the warmest 30-year period of the last 1,400 years (in the Northern Hemisphere, where assessment is possible). And all five of the years from 2014 to 2018 were the hottest on record globally. If warming trends continue at the current rate, it's estimated global warming will reach 2.7 degrees Fahrenheit (1.5 degrees Celsius) above preindustrial levels between 2030 and 2052.

Fueled by man-made greenhouse gas emissions, global warming is altering the earth's climate systems in many ways. It is:

- Causing more frequent and/or intense extreme weather events, including heat waves, hurricanes, droughts, and floods.
- Exacerbating precipitation extremes, making wet regions wetter and dry regions drier.
- Raising sea levels due to melting glaciers and sea ice and an increase in ocean temperatures (warmer water expands, which can contribute to sea level rise).

- Altering ecosystems and natural habitat, shifting the geographic ranges, seasonal activities, migration patterns, and abundance of land, freshwater, and marine species.

These changes pose threats not only to plants and wildlife, but directly to people. Warmer temperatures mean insects that spread diseases like dengue fever and Zika can thrive—and heat waves are getting hotter and more lethal to humans. People could go hungry when our food supply is diminished thanks to droughts and floods—a 2011 National Research Council study found that for every degree Celsius that the planet heats up, crop yields will go down 5 to 15 percent. Food insecurity can lead to mass human migration and political instability. And in January 2019, the Department of Defense released a report that described the threats to U.S. military installations and operations around the world due to flooding, droughts, and other impacts of climate change.

## The Greenhouse Effect Solution

The earth has always experienced warm and cool phases, with natural forces—from the sun's intensity, volcanic eruptions, and natural changes in greenhouse gas concentrations—affecting how much energy from the sun our planet absorbs. Scientists say that as recently as a couple of centuries ago, the planet underwent a "Little Ice Age," caused by a decrease in solar activity and an increase in volcanic activity. But today's climatic warming—particularly the increase in temperatures since the mid-20th century—is occurring at a pace that can't be explained by natural causes alone. According to NASA, "natural causes are still in play today, but their influence is too small or they occur too slowly to explain the rapid warming seen in recent decades."

In other words, humans are the problem. But we may also be the solution. We have the ability to rein in greenhouse gas emissions, though doing so certainly won't be easy. Overhauling our energy systems will require transformative, aggressive global action—and now. According to the IPCC, we must decrease greenhouse gas pollution by 45 percent from 2010 levels by 2030 and reach net zero emissions by 2050. To allow global warming to exceed 1.5 degrees Celsius (which the IPCC has identified as the threshold for avoiding climate change's worst impacts) would mean more intense drought, extreme heat, flooding, and poverty, the decline of species (including a mass die-off of the world's coral reefs), and the worsening of food shortages and wildfires.

Reducing our greenhouse gas emissions will require significant effort at the international, national, and local levels. First and foremost, we must slash fossil fuel production, consumption, and pollution by ramping up our use of clean, renewable energy and energy-efficient technologies and by investing in fuel-efficient and electric vehicles. We must end fossil fuel subsidies and better leverage "cap and invest" programs, carbon pricing, and carbon capture, storage, and utilization technologies (which catch the carbon dioxide from emissions sources like power plants or directly from the air and permanently bury it underground or convert it into other materials). We must protect our carbon-storing forests and reduce food waste and the emissions that go with it. And as individuals, we must commit to taking carbon-cutting actions in our daily lives.

Currently the United States faces the additional hurdle of an administration doubling down on fossil fuel use by rolling back standards aimed at reducing emissions from dirty power plants and cars and trucks (in other words, from the electricity and transportation sectors, the nation's two largest sources of greenhouse gas emissions). President Trump is also working to withdraw the nation from the landmark 2015 Paris climate agreement even though nearly two-thirds of Americans believe we should do more to tackle climate change, not less.

Still, decision makers, companies, leaders, and activists across the country and around the world staunchly believe we must act on climate change. For just as the emissions of man-made greenhouse gases long ago are inducing the climate change we see now, the emissions we release today will impact us long into the future.

## Burning Fossil Fuels Heats the Climate and Harms Public Health

### *Karin Kirk*

*Karin Kirk is a scientist and freelance writer whose work focuses on climate change and its cultural impacts.*

Air pollution kills people. This is not an abstract, faraway, or uncertain conclusion.

Public health researchers are beginning to conclude there is no safe level of air pollution. Even small doses trigger health problems. And the greater the concentration, the worse the health outcomes. All told, outdoor air pollution is among the world's greatest public health risks, responsible for nearly 4.5 million deaths worldwide in 2015.

Why is this relevant to climate change? The primary solution to climate change is also the most potent way to tackle air pollution: Burn less fossil fuel. Resistance remains to taking action to address climate change, particularly among Republican lawmakers, in part because some of the worst consequences of climate change seem far away in time and space. Some might see melting ice sheets, human migrations in other parts of the world, or species extinctions in boreal forests as unlikely or unimportant.

But shifting the topic to air pollution allows for a more direct way to connect the problem with its consequences.

In the U.S., air pollution kills around 100,000 people every year. It's the cause of 3% of all U.S. deaths, which is more deaths than traffic accidents and homicides combined, and air pollution costs the American economy up to $1 trillion per year. So while some still debate the greenhouse effect, few can deny the importance of saving American lives—and lungs.

"Burning Fossil Fuels Heats the Climate. It Also Harms Public Health," by Karin Kirk, Yale Climate Connections, March 13, 2020. Reprinted by permission.

## Why Air Pollution Is Deadly

The most damaging culprit is called PM 2.5—particulate matter less than 2.5 microns in size. About 200 PM 2.5 particles could fit end-to-end across a single grain of sugar. The human body is well equipped to deal with large airborne particles like pollen and heavy dust. But smaller particles are inhaled deeply into the lungs and can enter the bloodstream.

Health effects from fine particulate matter include aggravated asthma, respiratory infections, lung cancer, heart disease, stroke, cognitive impairment, and premature death. The World Health Organization has classified small particulates as a Group 1 carcinogen, which means there is sufficient evidence to conclude that it can cause cancer in humans.

Sources of human-caused particulate matter include fossil fuel combustion, vehicles, agriculture, road dust, and fires.

Small particulate matter is the most ravaging and deadly form of air pollution, but it's not the only kind. Sulfur dioxide, nitrogen oxides, volatile organic compounds, ozone, and ammonia also cause major health problems.

## How Air Pollution Affects People in the U.S.

When many people imagine unhealthy air, images of Beijing or Delhi come to mind. Indeed, air pollution kills millions of people in Asia and India. But the toll right here at home is far greater than many realize. A number of recent studies found that human-caused outdoor air pollution causes 100,000-200,000 premature U.S. deaths each year.

The geographic distribution of health problems varies with the specific source of pollution. Coal power plants diminish public health in the Appalachian region. Vehicle pollution is greatest in urban areas. And industrial pollution occurs in hotspots throughout Texas and the Southeast.

As is the case for many environmental problems, those who bear the gravest consequences are not the ones who cause the damage. In the U.S., areas with high poverty rates endure disproportionately high health and economic impacts from air pollution. Furthermore, a recent study found that non-Hispanic whites breathe in around 17% less air pollution than they cause by their own consumption, while black and Hispanic people inhale more than 50% more pollution than is generated by their actions.

## The Economic Cost of Air Pollution in the U.S.

Illness and loss of life exert a heavy burden to humanity, but the harm doesn't end there. Health problems inflict a financial toll because of health care costs, lost productivity from missed work days and school days, and reduced economic growth. The price tag for the health effects of human-caused air pollution in the U.S. is estimated at $886 billion to $1 trillion per year.

To look at it another way, the high financial consequence for air pollution offers a direct incentive to fix the problem. Money spent reducing pollution is money saved in preventing the damage. The Clean Air Act offers a stunning example: It cost $65 billion to implement and saved $2 trillion in avoided costs—a 30-fold payback.

Energy consumption accounts for the lion's share of human-caused air pollution, causing 57% of the economic damage, or $505 billion per year in the U.S. Included in that number are $248 billion in annual costs from transportation-related pollution, and a $124 billion price tag for pollution from electricity generation. These expenses are often considered an indirect subsidy for the fossil fuel industry because the industry levies the damage while taxpayers bear the consequences.

## Good News: Reduction in Coal Burning Saves Lives

Thanks in part to environmental regulations, the U.S. has had generally improving air quality over the past 50 years.

Further gains in public health have been made as economic headwinds have shuttered coal-burning power plants. A recent paper in the research journal *Nature Sustainability* estimates that 26,600 American lives have been saved between 2005 and

2016 because of coal plant closures, along with an additional benefit of saving 570 million bushels of corn, soybeans, and wheat near the power plants.

To put those promising results in perspective, coal-fired generation in the U.S. during that same time period caused 329,000 deaths and suppressed crop yields by 10.2 billion bushels. Although progress is being made, much work remains to be done.

## Bad News: Air Pollution Is Getting Worse and Regulations Are Being Weakened

Reducing air pollution is a goal shared by some but not all policymakers. In 2017, President Trump appointed contrarian toxicologist Robert Phalen to the EPA's Science Advisory Board, which reviews scientific information that is used to make or revise EPA regulations. Phalen has said that "Modern air is a little too clean for optimum health," a claim not supported by well-established science. At the Science Advisory Board, he joined others with industry-friendly interpretations of environmental priorities.

The EPA's "Affordable Clean Energy" rule would worsen air quality in the U.S., according to the EPA's own analysis. The proposed rules would result in 1,000 additional air pollution deaths per year by 2030, along with 48,000 cases of exacerbated asthma, 42,000 lost work days and 60,000 missed school days. By 2030 the annual price tag for increasing health impairments because of the relaxed rules could reach $11 billion.

Meanwhile, EPA data show an uptick in PM 2.5 since 2016. According to a study published by the National Bureau of Economic Research, the rise in particulate pollution was caused by growth in natural gas burning, increased vehicle use, lack of enforcement of air quality standards, and wildfires. PM 2.5 rose by 5.5% from 2016 to 2018, while causing 9,700 additional premature deaths in 2018 and inflicting economic damages of $89 billion.

## The Immediate Benefits of Addressing Air Pollution

Actions to reduce greenhouse gas emissions can be a hard sell among skeptical audiences. Common avenues of resistance are that it will cost too much, or that China is the real culprit of climate change, or that the effects of climate action are too intangible. But by shifting the topic to air pollution, those arguments may fall away. Reducing air pollution by burning less fossil fuels offers concrete, immediate, and local benefits for people and for the economy.

Case in point: A worldwide phase-out of fossil fuels would prevent the premature deaths of 3 to 4 million people each year, according to a 2019 study. Another study concluded that aggressive climate policy would avert 150 million air-pollution deaths worldwide over the next 80 years—and this is in addition to the social benefits of reducing the climate impacts like heat stress, flooding, and crop loss.

As policymakers pencil out the costs for climate mitigation, it's worth noting that the economic gains from reducing air pollution substantially outweigh the cost of achieving greenhouse gas emissions targets at either 1.5 degree Celsius or 2 degree Celsius thresholds. In some cases, the savings from avoiding pollution would be more than double the costs of emissions reduction, with the biggest benefits going to nations like China and India. In the U.S., an estimated 10 to 41% of the price tag for lowering greenhouse emissions could be offset by improvements in public health.

A rhetorical advantage to talking about air pollution is that unlike climate change, the issue dodges the steady drumbeat of dismissive talking points. While cries of, "But the climate has changed before!" resonate across any discussion of climate change, there are few arguments that can be levied in support of air pollution.

Better yet, a fresh angle on climate solutions can pave the way for building common ground. By sidestepping the land mines of tired talking points, it's easier to direct the discussion to where it's needed most: how to ease the burden of pollution on the planet and ourselves.

Current Controversies

# Chapter 4

# Are Climate Change and Loss of Biodiversity Preventable?

## Overview: All of Us Can Help Stop Global Warming

*Melissa Denchak*

*Melissa Denchak is a freelance editor and writer for the Natural Resources Defense Council.*

Nations around the world are upping their game in the fight against climate change, even as President Trump recently announced the U.S.'s withdrawal from the Paris Agreement. And despite this reckless move, American mayors, state leaders, county officials, governors, major companies, and millions of citizens across our country have pledged that they're "still in" when it comes to the agreement, and supporting the goal of limiting future warming to well below 2 degrees Celsius.

Even better, a new initiative by former New York City mayor Michael Bloomberg gives the urban layer of this movement a boost. He's asked mayors from the 100 most populous cities in the country to share their plans for making their buildings and transportation systems run cleaner and more efficiently. The 20 that show the greatest potential for cutting the dangerous carbon pollution that's driving climate change will share a total of $70 million in technical assistance funding provided by Bloomberg Philanthropies and partners.

It's important to remember the equally vital contributions that can be made by private citizens—which is to say, by you. "Change only happens when individuals take action," Aliya Haq, deputy director of NRDC's Clean Power Plan initiative, says. "There's no other way, if it doesn't start with people."

The goal is simple. Carbon dioxide is the climate's worst enemy. It's released when oil, coal, and other fossil fuels are burned for energy—the energy we use to power our homes, cars, and

"How You Can Stop Global Warming," by Melissa Denchak, NRDC (Natural Resources Defense Council), July 17, 2017. Reprinted by permission.

smartphones. By using less of it, we can curb our own contribution to climate change while also saving money. Here are a dozen easy, effective ways each one of us can make a difference:

## 1. Speak Up!

What's the single biggest way you can make an impact on global climate change? "Talk to your friends and family, and make sure your representatives are making good decisions," Haq says. By voicing your concerns—via social media or, better yet, directly to your elected officials—you send a message that you care about the warming world. Encourage Congress to enact new laws that limit carbon emissions and require polluters to pay for the emissions they produce. "The main reason elected officials do anything difficult is because their constituents make them," Haq says.

## 2. Power Your Home with Renewable Energy

Choose a utility company that generates at least half its power from wind or solar and has been certified by Green-e Energy, an organization that vets renewable energy options. If that isn't possible for you, take a look at your electric bill; many utilities now list other ways to support renewable sources on their monthly statements and websites.

## 3. Weatherize, Weatherize, Weatherize

"Building heating and cooling are among the biggest uses of energy," Haq says. Indeed, heating and air-conditioning account for almost half of home energy use. You can make your space more energy efficient by sealing drafts and ensuring it's adequately insulated. You can also claim federal tax credits for many energy-efficiency home improvements.

## 4. Invest in Energy-Efficient Appliances

Since they were first implemented nationally in 1987, efficiency standards for dozens of appliances and products have kept 2.3 billion tons of carbon dioxide out of the air. That's about the same amount as the annual carbon pollution coughed up by nearly 440 million cars. "Energy efficiency is the lowest-cost way to reduce emissions," Haq says. When shopping for refrigerators, washing machines, and other appliances, look for the Energy Star label. It will tell you which are the most efficient.

## 5. Reduce Water Waste

Saving water reduces carbon pollution, too. That's because it takes a lot of energy to pump, heat, and treat your water. So take shorter showers, turn off the tap while brushing your teeth, and switch to WaterSense-labeled fixtures and appliances. The EPA estimates that if just one out of every 100 American homes were retrofitted with water-efficient fixtures, about 100 million kilowatt-hours of electricity per year would be saved—avoiding 80,000 tons of global warming pollution.

## 6. Actually Eat the Food You Buy—And Make Less of It Meat

Approximately 10 percent of U.S. energy use goes into growing, processing, packaging, and shipping food—about 40 percent of which just winds up in the landfill. "If you're wasting less food, you're likely cutting down on energy consumption," Haq says. And since livestock products are among the most resource-intensive to produce, eating meat-free meals can make a big difference, too.

## 7. Buy Better Bulbs

LED lightbulbs use up to 80 percent less energy than conventional incandescents. They're also cheaper in the long run: A 10-watt LED that replaces your traditional 60-watt bulb will save you $125 over the lightbulb's life.

## 8. Pull the Plug(s)

Taken together, the outlets in your home are likely powering about 65 different devices—an average load for a home in the U.S. Audio and video devices, cordless vacuums and power tools, and other electronics use energy even when they're not charging. This "idle load" across all U.S. households adds up to the output of 50 large power plants in the U.S. So don't leave fully charged devices plugged into your home's outlets, unplug rarely used devices or plug them into power strips and timers, and adjust your computers and monitors to automatically power down to the lowest power mode when not in use.

## 9. Drive a Fuel-Efficient Vehicle

Gas-smart cars, such as hybrids and fully electric vehicles, save fuel and money. And once all cars and light trucks meet 2025's clean car standards, which means averaging 54.5 miles per gallon, they'll be a mainstay. For good reason: Relative to a national fleet of vehicles that averaged only 28.3 miles per gallon in 2011, Americans will spend $80 billion less at the pump each year and cut their automotive emissions by half.

## 10. Maintain Your Ride

If all Americans kept their tires properly inflated, we could save 1.2 billion gallons of gas each year. A simple tune-up can boost miles per gallon anywhere from 4 percent to 40 percent, and a new air filter can get you a 10 percent boost.

## 11. Rethink Planes, Trains, and Automobiles

Choosing to live in walkable smart-growth cities and towns with quality public transportation leads to less driving, less money spent on fuel, and less pollution in the air. Less frequent flying can make a big difference, too. "Air transport is a major source of climate pollution," Haq says. "If you can take a train instead, do that."

## 12. Shrink Your Carbon Profile

You can offset the carbon you produce by purchasing carbon offsets, which represent clean power that you can add to the nation's energy grid in place of power from fossil fuels. But not all carbon offset companies are alike. Do your homework to find the best supplier.

## Controlled Fires Maintain Habitat and Preserve Biodiversity

### *Forest Preserve District of Will County*

*The Forest Preserve District of Will County actively pursues a mission of protecting and conserving their natural areas.*

Fire is as natural a force as wind, water, drought, floods, blizzards, and insect infestations. Before Europeans settled in North America, fires regularly occurred naturally due to lightning strikes, but were also started accidentally and intentionally by man.

Deliberately set fires were an important tool of Native Americans who used fires to hunt, improve visibility, protect themselves and their villages from wildfires, make traveling through the tallgrass prairie easier, and for many other reasons.

As the continent was populated and developed, fire was widely suppressed because of its inherently destructive impact to many human interests and its potentially deadly affect on human life. The "Smoky the Bear" campaign reflects this perspective.

As a result, it is natural to have questions about the Forest Preserve District of Will County's prescribed burning practices, and we hope to answer those questions here.

### Why Is Fire Important?

In recent times, biologists have realized the benefits of fire in natural areas. Today we understand that our ecosystems and the plants within them evolved with fire, and many species are dependent on fire to maintain the habitat in which they live.

For example, the key growing part of most prairie plants is below ground, where the heat of the fire does not penetrate, allowing the grasses and wildflowers to flourish once again following a fire event. Many native trees such as oaks and hickories have evolved

"Prescribed Burning," The Forest Preserve District of Will County. Reprinted by permission.

adaptations to protect them from fire injuries such as thick bark. There are species of evergreen trees that cannot germinate until the cones are exposed to the heat generated by fire.

Historically, the landscape of Will County was dominated by prairie and wetlands, with scattered areas of timber that supported a wide variety of wildlife. While this landscape was molded through millennia by geology, topography, hydrology, and climatic factors, it was also strongly influenced by fire.

Explorers and early settlers remarked on the awesome power of the nearly annual prairie fires, and how in the absence of fire, trees became more common across the landscape. Fire so profoundly influenced the landscape that trees could only persist in areas sheltered from intense fire by natural firebreaks such as rivers and hills.

These firebreaks provide opportunities for the fire to go out or burn around and skip some areas. These skips and areas of reduced fire intensity allow saplings to grow large enough to withstand the fires and grow into the next generation of mature trees, sometimes forming woodlands or more sparsely timbered savanna communities.

In the absence of natural firebreaks where fire could burn most intensely, tree growth was prevented and prairie was maintained. The different plant species that dominated the prairie, woodland, or savanna landscapes provided the diversity of habitats which supported the large array of animal species we know today.

## What Is Prescribed Burning?

Prescribed (or "controlled") burning is a means of reintroducing this natural process. A controlled burn involves identifying the area to be burned (the "burn unit"), establishing control lines in order to prevent the fire from burning unintended areas, and intentionally setting the burn unit on fire.

Areas within the county are managed by the District with controlled burning when there are reasonable assumptions that wildfire was an important ecological process that shaped composition and structure of a particular ecosystem type, and that maintenance or restoration of that community type cannot be achieved without its use.

Fire can also be an important tool to attain specific maintenance objectives such as controlling tree and shrub growth along the embankment of a flood control reservoir where the roots systems can threaten structural integrity. The disturbance created by periodic fires is important in maintaining our remnant oak/hickory forests and savannas.

The Public Land Survey (1821-1837) recorded almost 1,500 trees while surveying Will County to establish the system of townships and sections that is still used today. Over 90% of those trees were oaks and hickories. Less than 1.5% of the trees recorded were ash, sugar maple, hackberry, or black cherry.

Now these once uncommon trees have rapidly colonized our landscape, threatening to overwhelm the oak/hickory woodlands and the plants and animals that have evolved with them. The one factor that these native invaders have in common is that their saplings are much more easily harmed by fire than those of oaks and hickories. Experiments have shown that reintroducing fire into oak/hickory ecosystems is an effective means of reestablishing oak and hickory sapling dominance.

Controlled burning is also an important tool providing many benefits for restoring our natural plant communities to lands that have been disturbed. Fire is used to clear the ground of existing vegetation in preparation of seeding and planting more desirable plant species. Burning allows the seeds to make better contact with the soil and therefore improves the chances of successful germination.

Fire is used to reduce the competition between weedy species and native species as a restoration area matures to encourage better establishment of the native vegetation which slowly displaces the weeds. Controlled burning also returns nutrients to the soil, making them readily available for the next generation of vegetation growth.

## How Is Prescribed Burning Done?

Controlled burning requires extensive planning, training, personnel, and equipment. Planning is often required six months or more prior to implementation.

An open-burn permit from the Illinois Environmental Protection Agency must be obtained. Adjacent landowners and the appropriate local Fire Chiefs are notified by mail well before the day of the burn. Dispatch centers, fire chiefs, and Forest Preserve District Police are also notified on the day of the burn.

All participants of our controlled burns have received special training, which involves learning fire behavior, methods of ignition and suppression, proper use of equipment, and more. The burn boss and some of the fire crew leaders also have extensive training in modeling fire behavior to assist with understanding expected and observed conditions.

A written plan is developed for every area to be managed with controlled burning. This burn plan identifies the limits of the burn unit and control line needs, potential hazards and safety issues, procedures for ignition, acceptable weather and fuel conditions, personnel and equipment needs, as well as the management goals and objectives of conducting the burn. Before every fire, burn plans are reviewed and updated as needed.

Fire control lines are installed prior to the day of the burn and checked immediately before ignition. If necessary to protect site visitors, access to portions of a preserve may be restricted or the preserve may be closed for the duration of the burn.

Data is recorded during and at the conclusion of each burn, including weather conditions, observed fire behavior, and extent of burn coverage, in order to help staff evaluate the effectiveness of the fire in achieving site management objectives, as well as safety and operational procedures. The controlled burn season spans mid-October to mid-April when most vegetation and many animals are dormant.

## How Are Injuries to Wildlife Avoided?

For those who have never seen the aftermath of a controlled burn, it may appear shocking at first. A blackened and smoking landscape greets the eye, and it appears as if all life has been destroyed. The reality is far different.

We carefully plan and time our burns to minimize harm to wildlife. Whenever practical, sites are divided into multiple burn units so that there is always unburned habitat within the preserve. While an occasional mouse or snake may be harmed in a burn, it is far more common to see a mouse, rabbit, or deer dart through the flames to safety than to see one actually harmed.

Controlled burn participants routinely walk through the burn units afterward to look for injured animals and have learned to burn in a manner that results in very few injuries or mortalities. Most insects are underground during controlled burns, but some species are in the plant litter and are thus consumed in the flames. However, leaving unburned areas ensures there are enough insects to accomplish the pollination and decomposition needs of the next growing season.

## Is It Dangerous?

Of course, any fire can be dangerous if not kept under control. Prior to a controlled burn, variables affecting fires are carefully studied. Our ecologists and fire control crews who are trained to meet National Wildfire Coordinating Group (NWCG) standards monitor wind conditions, humidity, temperature, and the amount of moisture in plant material.

Through training and experience, the controlled burn crews are able to anticipate problems and take remedial actions that have allowed this program to effectively manage our natural lands while maintaining public and burn participant safety for over 25 years without a single significant incident.

We strive to implement our controlled burn program with the greatest attention paid to the safety of the general public. An important element of this is smoke management. Common smoke management considerations include excessive smoke on roadways, which creates visibility concerns for motorists, and individuals with health conditions that can be aggravated by smoke such as asthma or allergies. Facilities with important fresh air intake systems like hospitals and schools are also vulnerable.

Several strategies exist for minimizing impacts resulting from smoke, including monitoring weather conditions which influence smoke dispersal, altering ignition strategies, and proper site selection the day of the burn.

We constantly update our burn plans to identify the location of these smoke sensitive receptors. If you or a family member has a health condition that could be triggered or aggravated by smoke, or could otherwise be negatively impacted by controlled burn activities, please contact us so that we can include your information in the appropriate burn plan(s). We will be sure to give you advance notification of planned controlled burn activities so that exposure to smoke or other impacts can be avoided.

## Can I Participate?

Yes. Anyone interested can assist District staff and participate in a controlled burn. Participants must first register as a Prairie People Volunteer with the District and complete the required safety and orientation training. Of course, basic controlled burn training is also required, but is provided by the District as is most necessary personal protection equipment. Once trained, volunteers may participate in burns as often as they wish.

# Prevent Climate Change and Loss of Biodiversity to Prevent Pandemics

***Harvard T. H. Chan School of Public Health***

*The Harvard T. H. Chan School of Public Health works to improve public health today and in the future for all people.*

Below are some of the most common questions we have been receiving in relation to the environment and coronavirus 2019 (COVID-19).

## Does Climate Change Affect the Transmission of Coronavirus?

We don't have direct evidence that climate change is influencing the spread of COVID-19, but we do know that climate change alters how we relate to other species on Earth and that matters to our health and our risk for infections.

As the planet heats up, animals big and small, on land and in the sea, are headed to the poles to get out of the heat. That means animals are coming into contact with other animals they normally wouldn't, and that creates an opportunity for pathogens to get into new hosts.

Many of the root causes of climate change also increase the risk of pandemics. Deforestation, which occurs mostly for agricultural purposes, is the largest cause of habitat loss worldwide. Loss of habitat forces animals to migrate and potentially contact other animals or people and share germs. Large livestock farms can also serve as a source for spillover of infections from animals to people. Less demand for animal meat and more sustainable animal husbandry could decrease emerging infectious disease risk and lower greenhouse gas emissions.

We have many reasons to take climate action to improve our health and reducing risks for infectious disease emergence is one of them.

## Does Air Pollution Increase the Risk of Getting Coronavirus? Does It Make Symptoms Worse?

Recent research from Rachel Nethery, Xiauo Wu, Francesca Dominici and other colleagues at Harvard Chan has found that people who live in places with poor air quality are more likely to die from COVID-19 even when accounting for other factors that may influence risk of death such as pre-existing medical conditions, socioeconomic status, and access to healthcare.

This finding is consistent with prior research that has shown that people who are exposed to more air pollution and who smoke fare worse with respiratory infections than those who are breathing cleaner air, and who don't smoke.

In places where air pollution is a routine problem, we have to pay particular attention to individuals who may be more exposed or vulnerable than others to polluted air, such as the homeless, those who don't have air filtration in their homes, or those whose health is already compromised. These individuals may need more attention and support than they did even before coronavirus came along.

For those interested in research papers on air pollution and virus transmission:

- Exposure to air pollution and COVID-19 mortality in the United States (Harvard University, preprint, 2019). This study found that a small increase in long-term exposure to PM2.5 leads to a large increase in COVID-19 death rate.
- Measuring the impact of air pollution on respiratory infection risk in China (*Environmental Pollution*, 2018). This study found that worse air quality in China may increase transmission of infections that cause influenza-like illnesses.
- The association between respiratory infection and air pollution in the setting of air quality policy and economic change (*Annals of the American Thoracic Society*, 2019). A

study of nearly 500,000 New York residents found that higher particulate matter air pollution levels increased the chances of hospitalization for pneumonia and emergency department visits, especially for influenza.

- Airborne transmission may have played a role in the spread of 2015 highly pathogenic avian influenza outbreaks in the United States (*Scientific Reports*, 2019). Researchers have found that several viruses, including adenovirus and influenza virus, can be carried on air particles. This recent paper finds that particulate matter likely contributed to the spread of the 2015 avian influenza.
- Relationship between ambient air pollution and daily mortality of SARS in Beijing (*Biomedical and Environmental Sciences*, 2005). During the SARS epidemic in 2003, this study found that increases in particulate matter air pollution increased risks of dying from the disease. SARS is a coronavirus, like COVID-19.

## Will Warmer Weather Slow the Spread of Coronavirus?

We don't yet have a sense of what the changing weather will mean for COVID-19 and so we shouldn't rely upon warmer weather to curtail transmissions. We need to do everything we can right now to slow the spread of this disease, and that means we need to follow the advice that public health experts are telling us and practice social distancing and good hand hygiene, among other actions.

## How Likely Are We to See Infectious Disease Spread as a Result of Climate Change?

Climate change has already made conditions more favorable to the spread of some infectious diseases, including Lyme disease, waterborne diseases such as Vibrio parahaemolyticus, which causes vomiting and diarrhea, and mosquito-borne diseases such as malaria and dengue fever. Future risks are not easy to foretell, but climate change hits hard on several fronts that matter to when and where pathogens appear, including temperature and rainfall patterns. To help limit the risk of infectious diseases, we should do all we can to vastly reduce greenhouse gas emissions and limit global warming to 1.5 degrees.

## Why Are Emerging Infectious Diseases on the Rise?

We have seen a trend of greater emergence of infectious diseases in recent decades. Most of these diseases have entered into people from animals, especially wild animals. This trend has many causes. We have massive concentrations of domesticated animals around the world, some of which can be home to pathogens, like the flu, that can make people sick. We also have massive concentrations of people in cities where diseases transmitted by sneezing may find fertile ground. And we have the ability to travel around the globe in less than a day and share germs widely.

But a look at the origins of COVID reveals that other forces may be in play. In the past century we have escalated our demands upon nature, such that today, we are losing species at a rate unknown since the dinosaurs, along with half of life on earth, went extinct 65 million years ago. This rapid dismantling of life on earth owes primarily to habitat loss, which occurs mostly from growing crops and raising livestock for people. With fewer places to live and fewer food sources to feed on, animals find food and shelter where people are, and that can lead to disease spread.

Another major cause of species loss is climate change, which can also change where animals and plants live and affect where diseases may occur. Historically, we have grown as a species in partnership with the plants and animals we live with. So, when we change the rules of the game by drastically changing the climate and life on earth, we have to expect that it will affect our health.

## What Actions Can We Take to Prevent Future Outbreaks?

We can make many smart investments to avert another outbreak. Federal, state, and local agencies can support public health leadership and science, we can provide more funding for needed research, early response to outbreaks, and supplies for testing. And we can do much more to control the illegal wildlife trade.

We also need to take climate action to prevent the next pandemic. For example, preventing deforestation—a root cause of climate change—can help stem biodiversity loss as well as slow animal migrations that can increase risk of infectious disease spread. The recent Ebola epidemic in West Africa probably occurred in part because bats, which carried the disease, had been forced to move into new habitats because the forests they used to live in had been cut down to grow palm oil trees.

Rethinking our agricultural practices, including those that rely on raising tens of millions of animals in close quarters, can prevent transmissions between animals and spillover into human populations.

Reducing air pollution caused by burning fossil fuels like coal, oil and natural gas also helps keep our lungs healthy, which can protect us from respiratory infections like coronavirus.

To combat climate change, we need to drastically decrease greenhouse gas emissions. Generating electricity from low-carbon energy sources like wind and solar decreases harmful air pollutants such as nitrogen oxides, sulfur dioxide, and carbon dioxide that lead to more heart attacks and stroke as well as obesity, diabetes, and premature deaths that put further strains on our health care systems.

Preparation for pandemics is also about keeping people healthy at baseline. If we have a population in the U.S. where a third of our population are obese, and 5-10% of people have diabetes, we're going to be immensely more vulnerable. And if you look at why people in the U.S. are not healthy at baseline, it has to do with our diets, pollution, and climate change. We have an opportunity here to recognize that prevention is by far the best approach to protecting health.

When COVID-19 eases, and we are ready to restart our economy, we can make our workforce healthier and more climate-resilient through scaling-up our investments in low-carbon technologies.

## Can You Identify the Communities Most At-Risk, and How and Why Both COVID-19 and Climate Change Harm Them?

People with chronic health conditions, lower-income, and communities of color are disproportionately impacted by both COVID-19 and climate change, and pollution is at the heart of both problems as a new Harvard T. H. Chan School of Public Health study confirms. We know that African American communities are disproportionately exposed to air pollution and we're now seeing this pollution driving higher mortality rates from COVID-19. We owe it to everyone to improve health, and we do that by reducing the sources of pollution that drive a large burden of disease both in the United States and around the world.

## Why Is It So Important for Health Officials to Talk About Climate Change Now?

Having taken care of children and families who are deeply concerned about how they can protect their children from this disease, I can tell you that we need to wash our hands and we need to socially distance. But if we really care about preventing this kind of problem in the future, we need to think hard about climate change and the biodiversity crisis. I was actually in a room with a child and a family when I first thought that this is exactly the time that we need to think more about the broader issues that we face. We simply cannot afford to deal with a crisis like this pandemic on top of another climate-related crisis—like a hurricane, tornado, wildfire, or heatwave—when we absolutely know how to implement climate solutions, and can put them into action right now. Doing so will make us healthier today and protect our future.

## Climate Change and Global Health Policy Are Largely Treated as Separate Issues by the Public and Media. Do We Need to Adjust Our Thinking?

Yes. The separation of health and environmental policy is a dangerous delusion. Our health entirely depends on the climate and the other organisms we share the planet with. We need to bring these communities together. Some progress has been made in addressing the risk of pathogen spillover from animals into people. But largely we still view the environment, and life on earth, as separate. We can and must do better if we want to prevent the next infectious pandemic. That means we must combat climate change and do far more to safeguard the diversity of life on earth, which is being lost at a rate not seen since the dinosaurs—and more than half of life on earth—went extinct 65 million years ago.

## COVID-19 Is Killing People Now and Climate Change Is Killing People Now. The Scale of Actions to Combat Them Are Starkly Different. Why?

Infectious diseases are scary because they are immediate and personal. They radically and rapidly change how we lead our lives, and they are an immediate threat to our friends and families. They hit all of our "go" buttons.

Climate change seems to many an armageddon in slow motion and its dangers can feel impersonal and its causes diffuse. It's easy to think "I didn't cause this" or that "it doesn't directly affect me." But there's another way to look at it. Like COVID-19, if you're concerned about climate change, you can take actions right now to improve your health and the health of your friends and loved ones.

We can learn from this pandemic that people are motivated by the personal and the actionable. At Harvard Chan C-CHANGE, our research shows that the actions we need to combat climate change are the same actions we need to make people healthier right now, especially for diseases causing huge burdens on our health like obesity, heart disease, and cancer. We need to do much more to talk about the "burden of disease" that's preventable, and the things we can do now to prevent it.

## Is Climate Change Too Expensive to Fix?

We spend just over $3 trillion every year in the United States on health care. And by some estimates, more than half the deaths in the United States are preventable, largely because of pollution, diet, exercise, and lifestyle habits like smoking. So think about the money we could save simply by reducing air pollution, eating less meat, and building exercise into our day by walking or biking more often. We could use the savings to invest in preventing climate change, among other things like education, and paying fair wages.

When you look at this question purely from a financial standpoint, air pollution is a drag on economic growth and solutions to address have been enormously cost-effective in the United States. In 2011, a study by the Environmental Protection Agency that looked at the costs and benefits of the Clean Air Act found that every $1 invested to reduce air pollution returns up to $30 in benefits. The only thing our health and our economy can't afford is climate inaction.

# De-extinction: Technique and Strategy to Save Species

*Abby Norman*

*Abby Norman is a staff writer for* Futurism.

This week scientists at Harvard University announced that they had assembled a nearly complete genome of the little bush moa. Never spotted one of these at your backyard feeder? Well, neither has anyone else alive today: the flightless bird has been extinct for about 700 years.

But if scientists get their way, that could soon change. Assembling the complete genome of the moa—and those of other long-gone creatures like the woolly mammoth, passenger pigeons, or even our Neanderthal cousins—is the first step in bringing those species back to life. Or at least, scientists' closest approximation.

The researchers first got their hands on the moa DNA by extracting it from a big toe of a museum specimen, they told Stat News.

Using modern technology to revive extinct species is known simply as "de-extinction," though the process is not what you'd call simple.

It's like putting together a jigzaw puzzle that has 900 million pieces. And is still missing some. And you don't have the picture on the box. But you have a puzzle that's similar as reference.

The key components of all species' DNA are nucleotides: adenine, cytosine, guanine, and thymine—noted by the letters A, C, G, and T, respectively. The Harvard team started with 900 million of the moa's nucleotides to start piecing together its entire genome (they aren't saying how many they needed for a complete genome). Using the genome of a modern bird relative, the emu, to fill in the blanks, the researchers were able to piece

"3 Reasons Why It's a Good Idea to Resurrect a Species," by Abby Norman, *Futurism*. Reprinted by permission.

together about 85 percent of the bird's genome—on par with the genomes of other extinct species.

Once scientists organize the genome, the next step is deciding how to best attempt to revive the species. Researchers can opt for "backbreeding," a process in which they take a closely related species and selectively breed it until they produce something akin to the extinct species. If the species hasn't been extinct for very long and researchers have a sample from, perhaps, the last living member, cloning may also be a viable option. The third option—using genetic engineering—is relatively new. With an accurate assembly of an extinct species' genetic data, researchers can inject them into an egg of a living species that's genetically close to the one they're attempting to bring back to life.

So bringing species back from the dead is feasible, if improbable and resource-intensive. But the question remains: why revive extinct species at all? If we've learned anything from *Jurassic Park* (which, let's be honest, we all have) it's that "just because we can, doesn't mean we should." Here are some reasons that scientists have used to justify their work.

## To Improve Our Science

Every time scientists make sense of an extinct species' genetic information, they get better at doing it. Yes, the human researchers improve, but so do the computer algorithms that help them parse out the data. With each attempt, researchers learn more about the sequencing process, and can train the technology to perform its tasks more accurately and efficiently. And the genomes researchers are reconstructing now could be useful for assembling the DNA of other species in the future. Little by little, they're assembling a genetic Rolodex for nearly every species on Earth.

Science for science's sake is a noble effort, but there are also many arguments for "resurrection biology," which frame the effort as being less about the de-extinction of species that have been lost to us, and more about preventing the extinction of endangered species.

## To Save Endangered Species

As humans have taken up more space and resources, we've destroyed natural habitats and endangered species. Many of those species eventually go extinct, pushing biodiversity levels below the "safe" threshold—the level at which the ecosystems can regenerate and sustain us—around the globe. The less biodiversity in an ecosystem, the greater chance species after species will go extinct.

We'll feel the effects of biodiversity loss, especially when it comes to plants. The medications that we take are produced from or with plant extracts; the textiles and fabrics that we use in our clothing, and much of what we eat, all come from plant species. Plants depend on the animals and insects in their environments for pollination and propagation. In return, those critters rely on crops for nutrition and housing just as we do. If those pollinators go extinct, the plants—and the species, like us, that rely on those plants—also go extinct.

Some species, called keystone species, are more important than others in maintaining that biodiversity. The passenger pigeon, for example, (a relatively recent extinction and thus a good candidate for resurrection) shaped many of North America's forests. When the species was lost, the forests lost the main driver of their regulation cycle and have never been the same.

Reviving extinct keystone species, then, could help us preserve biodiversity, and, possibly, the ecosystems as a whole.

## To Preserve the Planet

Bringing now-dead species back to life might help Earth survive in the long run. At least, that's what Beth Shapiro, a biologist at the University of California, Santa Cruz, thinks. Shapiro has worked on several de-extinction projects (including the moa's). In a 2015 interview with *Smithsonian Magazine*, Shapiro said she believed that land conservation efforts are simply not enough to preserve the planet. Much of the world's biodiversity has been lost as a result of human interference: either directly via practices like deforestation, or indirectly, through our refusal to accept the realities of climate change.

If we want to get an idea of what happens to the world's ecosystems after a major climate shift, we need not look further than previous extinction events. The woolly mammoth, for example, used to fertilize the grasslands of Siberia; after it went extinct, the entire region became icy and barren.

"De-extinction may not be the answer to the biodiversity crisis that we are facing today, but the technologies that are being developed in the name of de-extinction may become powerful new tools in an active conservation regime," Shapiro told *Smithsonian*. "Why not provide populations a little bit of genomic assistance so they can survive in a world that is changing too quickly for natural evolutionary processes to keep up?"

Considering that humans have, for millennia, pushed out species that were here long before us… that seems like the very least we could do.

# Is the Sixth Mass Extinction Already Here?

***Frédérik Saltré and Corey J. A. Bradshaw***

*Frédérik Saltré and Corey J. A. Bradshaw are researchers in ecological studies for the ARC Centre of Excellence for Australian Biodiversity and Heritage.*

For more than 3.5 billion years, living organisms have thrived, multiplied and diversified to occupy every ecosystem on Earth. The flip side to this explosion of new species is that species extinctions have also always been part of the evolutionary life cycle.

But these two processes are not always in step. When the loss of species rapidly outpaces the formation of new species, this balance can be tipped enough to elicit what are known as "mass extinction" events.

A mass extinction is usually defined as a loss of about three quarters of all species in existence across the entire Earth over a "short" geological period of time. Given the vast amount of time since life first evolved on the planet, "short" is defined as anything less than 2.8 million years.

Since at least the Cambrian period that began around 540 million years ago when the diversity of life first exploded into a vast array of forms, only five extinction events have definitively met these mass-extinction criteria.

These so-called "Big Five" have become part of the scientific benchmark to determine whether human beings have today created the conditions for a sixth mass extinction.

## The Big Five

These five mass extinctions have happened on average every 100 million years or so since the Cambrian, although there is no detectable pattern in their particular timing. Each event itself lasted

"What Is a 'Mass Extinction' and Are We in One Now?" by Frédérik Saltré and Corey J. A. Bradshaw, The Conversation, November 12, 2019. https://theconversation.com/what-is-a-mass-extinction-and-are-we-in-one-now-122535. 

between 50 thousand and 2.76 million years. The first mass extinction happened at the end of the Ordovician period about 443 million years ago and wiped out over 85% of all species.

The Ordovician event seems to have been the result of two climate phenomena. First, a planetary-scale period of glaciation (a global-scale "ice age"), then a rapid warming period.

The second mass extinction occurred during the Late Devonian period around 374 million years ago. This affected around 75% of all species, most of which were bottom-dwelling invertebrates in tropical seas at that time.

This period in Earth's past was characterised by high variation in sea levels, and rapidly alternating conditions of global cooling and warming. It was also the time when plants were starting to take over dry land, and there was a drop in global $CO_2$ concentration; all this was accompanied by soil transformation and periods of low oxygen.

The third and most devastating of the Big Five occurred at the end of the Permian period around 250 million years ago. This wiped out more than 95% of all species in existence at the time.

Some of the suggested causes include an asteroid impact that filled the air with pulverised particles, creating unfavourable climate conditions for many species. These could have blocked the sun and generated intense acid rains. Some other possible causes are still debated, such as massive volcanic activity in what is today Siberia, increasing ocean toxicity caused by an increase in atmospheric $CO_2$, or the spread of oxygen-poor water in the deep ocean.

Fifty million years after the great Permian extinction, about 80% of the world's species again went extinct during the Triassic event. This was possibly caused by some colossal geological activity in what is today the Atlantic Ocean that would have elevated atmospheric $CO_2$ concentrations, increased global temperatures, and acidified oceans.

The last and probably most well-known of the mass-extinction events happened during the Cretaceous period, when an estimated 76% of all species went extinct, including the non-avian dinosaurs. The demise of the dinosaur super predators gave mammals a new opportunity to diversify and occupy new habitats, from which human beings eventually evolved.

The most likely cause of the Cretaceous mass extinction was an extraterrestrial impact in the Yucatán of modern-day Mexico, a massive volcanic eruption in the Deccan Province of modern-day west-central India, or both in combination.

## Is Today's Biodiversity Crisis a Sixth Mass Extinction?

The Earth is currently experiencing an extinction crisis largely due to the exploitation of the planet by people. But whether this constitutes a sixth mass extinction depends on whether today's extinction rate is greater than the "normal" or "background" rate that occurs between mass extinctions.

This background rate indicates how fast species would be expected to disappear in absence of human endeavour, and it's mostly measured using the fossil record to count how many species died out between mass extinction events.

The most accepted background rate estimated from the fossil record gives an average lifespan of about one million years for a species, or one species extinction per million species-years. But this estimated rate is highly uncertain, ranging between 0.1 and 2.0 extinctions per million species-years. Whether we are now indeed in a sixth mass extinction depends to some extent on the true value of this rate. Otherwise, it's difficult to compare Earth's situation today with the past.

In contrast to the Big Five, today's species losses are driven by a mix of direct and indirect human activities, such as the destruction and fragmentation of habitats, direct exploitation like fishing and hunting, chemical pollution, invasive species, and human-caused global warming.

If we use the same approach to estimate today's extinctions per million species-years, we come up with a rate that is between ten and 10,000 times higher than the background rate.

Even considering a conservative background rate of two extinctions per million species-years, the number of species that have gone extinct in the last century would have otherwise taken between 800 and

10,000 years to disappear if they were merely succumbing to the expected extinctions that happen at random. This alone supports the notion that the Earth is at least experiencing many more extinctions than expected from the background rate.

It would likely take several millions of years of normal evolutionary diversification to "restore" the Earth's species to what they were prior to human beings rapidly changing the planet. Among land vertebrates (species with an internal skeleton), 322 species have been recorded going extinct since the year 1500, or about 1.2 species going extinction every two years.

If this doesn't sound like much, it's important to remember extinction is always preceded by a loss in population abundance and shrinking distributions. Based on the number of decreasing vertebrate species listed in the International Union for Conservation of Nature's Red List of Threatened Species, 32% of all known species across all ecosystems and groups are decreasing in abundance and range. In fact, the Earth has lost about 60% of all vertebrate individuals since 1970.

Australia has one of the worst recent extinction records of any continent, with more than 100 species of vertebrates going extinct since the first people arrived over 50 thousand years ago. And more than 300 animal and 1,000 plant species are now considered threatened with imminent extinction.

Although biologists are still debating how much the current extinction rate exceeds the background rate, even the most conservative estimates reveal an exceptionally rapid loss of biodiversity typical of a mass extinction event.

In fact, some studies show that the interacting conditions experienced today, such as accelerated climate change, changing atmospheric composition caused by human industry, and abnormal ecological stresses arising from human consumption of resources, define a perfect storm for extinctions. All these conditions together indicate that a sixth mass extinction is already well under way.

# Climate Change Is On Track to Cause Abrupt Biodiversity Losses This Century

***Christopher Trisos and Alex Pigot***

*Christopher Trisos is a research fellow at University College London. Alex Pigot is a research fellow at the University of Cape Town in South Africa.*

The impacts of climate change on species and ecosystems are already evident. Poleward shifts in the geographic distributions of species, catastrophic forest fires and mass bleaching of coral reefs all bear the fingerprints of climate change.

But what will the world's biodiversity look like in the future?

Projections indicate that unless emissions are rapidly reduced the climate crisis will get substantially worse. Up to 50% of species are forecast to lose most of their suitable climate conditions by 2100 under the highest greenhouse gas emissions scenario.

But we still lack answers to some basic questions. When will species be exposed to potentially dangerous climate conditions? Will this occur in the next decade or only later in the century? Will the exposure of species accumulate gradually, one species at a time? Or should we expect abrupt jumps as the climate limits of multiple species are exceeded?

Our understanding of when and how abruptly climate driven disruptions of biodiversity will occur is limited because biodiversity forecasts typically focus on individual snapshots of the future. We took a different route. We used annual projections of temperature and precipitation from 1850 to 2100 across more than 30,000 marine and terrestrial species to estimate the timing of species exposure to potentially dangerous climate conditions.

Based on these projections, we estimate that climate change could cause sudden biodiversity losses. These could occur much sooner this century than had been expected. This new analysis indicates that a high percentage of species in local ecosystems could be exposed to potentially dangerous climate conditions simultaneously.

Rather than slowly sliding down a climate change slope, many ecosystems face a cliff edge.

## Risk of Abrupt Biodiversity Loss Early This Century

Abrupt biodiversity loss due to marine heatwaves that bleach coral reefs is already under way in tropical oceans. The risk of climate change causing sudden collapses of ocean ecosystems is projected to escalate further in the 2030s and 2040s. Under a high greenhouse gas emissions scenario the risk of abrupt biodiversity loss is projected to spread onto land, affecting tropical forests and more temperate ecosystems by the 2050s.

These dire projections use historical temperature models to find the upper limit that each species can survive under, as far as we know. Once temperatures rise to levels a species has never experienced, scientists have very limited evidence of their ability to survive.

It's possible some species, such as those with very short generation times, may be able to adapt. For species with longer generation times—such as most birds and mammals—it may be only a few generations before unprecedented temperatures occur. When this happens the species' ability to evolve out of this problem may be limited.

## Why It Matters

Abrupt losses of biodiversity from climate change represent a significant threat to human well-being. In many countries a large percentage of people rely on their immediate natural environment for their food security and income. Sudden disruption of local ecosystems would negatively affect their ability to earn an income and feed themselves, potentially pushing them into poverty.

For instance, marine ecosystems in the Indo-Pacific, Caribbean and the west coast of Africa are at high risk of sudden disruption as early as the 2030s. Hundreds of millions of people across these regions rely on wild-caught fish as an essential source of food. Eco-tourism revenues from coral reefs are also a major source of income.

In Latin America, Asia and Africa, large parts of the Andes, Amazon, Indonesian and Congo forests are projected to be at risk from 2050 under a high emissions scenario.

Sudden loss of animal communities could negatively affect the food security of people in these regions. It could also reduce the long-term ability of tropical forests to lock up carbon if the birds and mammals that are important for dispersing seeds are lost.

## Urgent Next Steps

These findings highlight the urgent need for climate change mitigation. Rapidly reducing greenhouse gas emissions this decade will help save thousands of species from extinction, and protect the life-giving benefits they provide to humans.

Keeping global warming below 2°C flattens the curve of climate change risk to biodiversity. It does this by massively reducing the number of species at risk and buys more time for species and ecosystems to adapt to the changing climate—whether that's by finding new habitats, changing their behaviour, or with the help of human-led conservation efforts.

There's also an urgent need to ramp up efforts to help people in high risk regions adapt their livelihoods as climate change alters local ecosystems.

Projecting where and when species will be exposed to dangerous climate change throughout the century could provide an early warning system, identifying those areas most at risk of abrupt ecological disruption. In addition to highlighting the urgent need for reducing fossil fuel usage, these results could help guide conservation efforts, such as designating new protected areas in climate refugia.

They could also inform resilient ecosystem-based approaches for helping people adapt to changing climates. An example would be planting mangroves to protect coastal communities against increasing flooding. The potential to continuously update and validate these near-term projections as ecological responses to climate change unfold should further refine projections of future climate risks to biodiversity that are so central to managing the climate crisis.

Our planet is still teeming with life. And with the right political leadership and daily actions that we take as citizens, we still have the power to keep it that way.

# Organizations to Contact

*The editors have compiled the following list of organizations concerned with the issues debated in this book. The descriptions are derived from materials provided by the organizations. All have publications or information available for interested readers. This list was compiled on the date of publication of the present volume; the information provided here may change. Be aware that many organizations take several weeks or longer to respond to inquiries, so allow as much time as possible.*

### American Action Forum-1

1747 Pennsylvania Avenue NW, 5th Floor
Washington, DC 20006
(202) 559-6420
email: contact@americanactionforum.org
website: www.americanactionforum.org

The American Action Forum was founded in 2009 as a nonprofit organization dedicated to educating the public about complex issues that the US needs to face. Read all about current proposals about climate change and proposed regulation and legislation.

### Citizen's Climate Lobby

1330 Orange Avenue, #309
Coronado, CA 92118
(619) 437-7142
email: use contact page
website: www.citizensclimatelobby.org

Citizen's Climate Lobby is an organization that empowers everyday people to work together on climate issues. You can join one of the over 500 local chapters around the US to do your part to change the climate crisis. Read, watch videos, and join up at this site.

## Climate Central

One Palmer Square, Suite 402
Princeton, NJ 08542
(877) 425-4724
email: use contact page
website: www.climatecentral.org

Climate Central is an organization made up of experts in science and journalism. They research, publish, and report on topics to keep interested citizens informed. Read, watch videos, and subscribe to the newsletter to keep up-to-date on the latest scientific findings about climate.

## Environmental Defense Fund (EDF)

1875 Connecticut Avenue NW, Suite 600
Washington, DC 20009
(202) 572-3298
email: use contact page
website: www.edf.org

The EDF uses a broad-based partner support network to enact global solutions for climate, energy, ecosystems, and other areas. Read about environmental success stories like saving rainforests or cleaner vehicles. Also sign up for news and alerts to get the latest environmental information.

## International Union for Conservation of Nature

1630 Connecticut Avenue NW, Suite 300
Washington, DC 20009
(202) 387-4826
website: www.iucn.org

This global organization is comprised of member governments, scientific experts, and interested citizens and organizations. Its aim is to conserve nature and lead the way to a sustainable future for the world. Learn about biodiversity and nature around the globe from an expert, reliable source.

### Meat Free Monday

email: info@meatfreemondays.com
website: www.meatfreemondays.com

Here's an easy way to make a difference for our climate. Find out how you can lower your consumption of meat, why that matters for climate change, and get recipes to be on your way.

### National Aeronautics and Space Administration (NASA)

300 E Street SW, Suite 5R30
Washington, DC 20546
(202) 358-0001
email: use links on contact page
website: www.nasa.gov

Most people are familiar with NASA's space technology, but it also specializes in climate science. Engage your mind in a vast source of exciting topics about weather and climate.

### Natural Resources Defense Council (NRDC)

40 W. 20th Street, 11th Floor
New York, NY 10011
(212) 727-2700
email: nrdcinfo@nrdc.org
website: www.nrdc.org

The NRDC works to safeguard the Earth—the people, animals, plants, and natural systems that work together so all can live. Learn about pending legislation and how to take action, such as writing an email to the president or your senator showing your concerns about the natural world.

### NOAA Climate

email: use contact page
website: www.climate.gov

This site has maps, videos, and articles covering the latest information about climate conditions across the US and the world. Track temperatures, precipitation, even glaciers on the extensive site.

### World Wildlife Fund

1250 24th Street NW
Washington, DC 20037
(202) 293-4800
email: membership@wwfus.org
website: www.worldwildlife.org

The WWF works to preserve nature and to lessen the threats to biodiversity around the globe. Its work centers on six areas—freshwater, food, climate, oceans, forests, and wildlife. Use this site for trustworthy scientific information about climate and the world's biodiversity.

# Bibliography

## Books

Stephen Aitken. *Ecosystems at Risk.* New York, NY: Cavendish Square, 2014.

Jeff Fleischer. *A Hot Mess: How the Climate Crisis Is Changing Our World.* Minneapolis, MN: Lerner Publishing, 2021.

Jane Goodall. *The Book of Hope: A Survival Guide for Trying Times.* New York, NY: Celadon Books, 2021.

Pamela Hill. *Environmental Protection: What Everyone Needs to Know.* New York, NY: Oxford University Press, 2017.

Marc Horst. *Palm Trees at the North Pole: The Hot Truth About Climate Change.* Vancouver, BC: Greystone Books, 2021.

Stuart A. Kallen. *Extreme Weather and Climate Change: What's the Connection?* San Diego, CA: Reference Point Press, 2022.

Stuart A. Kallen. *Stand Up Against Climate Change.* San Diego, CA: Reference Point Press, 2022.

Stuart A. Kallen. *What Is the Impact of Declining Biodiversity.* San Diego, CA: Reference Point Press, 2021.

Christina Limpert. *Climate: Powerful Voices, Inspiring Ideas.* Seattle, WA: Spruce Books, 2021.

Stephanie Sammartino McPherson. *Hothouse Earth: The Climate Crisis and the Importance of Carbon Neutrality.* Minneapolis, MN: Twenty-First Century Books, 2021.

Don Nardo. *Planet Under Siege: Climate Change.* San Diego, CA: Reference Point Press, 2020.

Sally Nex. *How to Garden the Low Carbon Way: The Steps You Can Take to Help Combat Climate Change.* New York, NY: DK, 2021.

Laura Perdew. *Biodiversity: Explore the Diversity of Life on Earth: With Environmental Science Activities for Kids.* White River Junction, VT: Nomad Press, 2019.

Jennifer Swanson. *Geoengineering Earth's Climate: Resetting the Thermostat.* Minneapolis, MN: Twenty-First Century Books, 2018.

John Woodward. *Climate Change.* New York, NY: DK Publishing, 2021.

## Periodicals and Internet Sources

Helen Briggs, "Climate Change: Polar Bears Could Be Lost by 2100," BBC, July 20, 2020. https://www.bbc.com/news/science-environment-53474445.

Catherine Clifford, "How You Can Help Fight Climate Change in Ways That Really Matter," CNBC, September 4, 2021. https://www.cnbc.com/2021/09/04/how-to-fight-climate-change-six-actions-that-really-matter.html.

Daniel W. Drezner, "Climate Change Is Real, and It's Permanent," *Washington Post*, August 10, 2021. https://www.washingtonpost.com/outlook/2021/08/10/climate-change-is-real-its-permanent/.

Saul Griffith, "We Can Beat Climate Change if We Do One Thing Fast," *Time*, November 4, 2021. https://time.com/6113719/optimistic-fighting-climate-change/.

Kanupriya Kapoor, "Climate Change and Biodiversity Loss Must Be Tackled Together—Report," Reuters, June 10, 2021. https://www.reuters.com/business/environment/climate-change-biodiversity-loss-must-be-tackled-together-report-2021-06-10/.

Paul Mann, "Could Resurrecting Mammoths Help Stop Arctic Emissions?" The Conversation, May 11, 2018. https://

theconversation.com/could-resurrecting-mammoths-help-stop-arctic-emissions-95956?xid=PS_smithsonian.

Matt McGrath, "Climate Change: Extreme Weather Events Are the 'New Norm,'" BBC, October 31, 2021. https://www.bbc.com/news/science-environment-59105963.

Jerry Melillo, "Forests and Climate Change," MIT, October 7, 2021. https://climate.mit.edu/explainers/forests-and-climate-change.

Razan Al Mubarak, "Can the Twin Crises of Biodiversity Loss and Climate Change Be Tackled Together?" World Economic Forum, May 26, 2021. https://www.weforum.org/agenda/2021/05/twin-crises-biodiversity-loss-and-mitigating-climate-change-tackled-together/.

Neil Ever Osbourne, "Polar Bears Live on the Edge of the Climate Change Crisis," *Smithsonian*, March 2021. https://www.smithsonianmag.com/science-nature/polar-bears-live-cliate-change-crisis-180976995/.

Gabriel Popkin, "How Much Can Forests Fight Climate Change?" *Nature*, January 15, 2019. https://www.nature.com/articles/d41586-019-00122-z.

Doyle Rice, "One-Third of All Plant and Animal Species Could Be Extinct in 50 Years, Study Says," *USA Today*, February 14, 2020. https://www.usatoday.com/story/news/nation/2020/02/14/climate-change-study-plant-animal-extinction/4760646002/.

Julia Rosen, "The Science of Climate Change Explained: Facts, Evidence and Proof," *New York Times*, April 19, 2021. https://www.nytimes.com/article/climate-change-global-warming-faq.html.

Ashley Strickland, "Modern Elephants Survived Ancient Climate Change. More Than 180 Other Species, Including Mastodons, Didn't," CNN, July 2, 2021. https://www.cnn.com/2021/07/02/world/mastodons-elephants-climate-change-scn/index.html.

# Index

## H



## I



## J



## K



## L



## M



## N

## O



## P



## R



## S



## T



## U



## V



## W